Lean-Six Sigma for the Public Sector

Also available from ASQ Quality Press:

Making Government Great Again: Mapping the Road to Success with ISO 9001:2008
John D. Baranzelli, P.E.

Office Kaizen 2: Harnessing Leadership, Organizations, People, and Tools for Office Excellence
William Lareau

Lean for Service Organizations and Offices: A Holistic Approach for Achieving Operational Excellence and Improvements
Debashis Sarkar

The Executive Guide to Understanding and Implementing Lean Six Sigma: The Financial Impact
Robert M. Meisel, Steven J. Babb, Steven F. Marsh, & James P. Schlichting

Six Sigma for the New Millennium: A CSSBB Guidebook, Second Edition
Kim H. Pries

5S for Service Organizations and Offices: A Lean Look at Improvements
Sarkar, Debashis

The Certified Six Sigma Black Belt Handbook, Second Edition
T.M. Kubiak and Donald W. Benbow

The Certified Six Sigma Green Belt Handbook
Roderick A. Munro, Matthew J. Maio, Mohamed B. Nawaz, Govindarajan Ramu, and Daniel J. Zrymiak

The Certified Six Sigma Master Black Belt Handbook
T.M. Kubiak

The Quality Toolbox, Second Edition
Nancy R. Tague

Mapping Work Processes, Second Edition
Bjørn Andersen, Tom Fagerhaug, Bjørnar Henriksen, and Lars E. Onsøyen

Root Cause Analysis: Simplified Tools and Techniques, Second Edition
Bjørn Andersen and Tom Fagerhaug

Lean ISO 9001: Adding Spark to your ISO 9001 QMS and Sustainability to your Lean Efforts
Mike Micklewright

Root Cause Analysis: The Core of Problem Solving and Corrective Action
Duke Okes

To request a complimentary catalog of ASQ Quality Press publications, call 800-248-1946, or visit our Web site at http://www.asq.org/quality-press.

Lean-Six Sigma for the Public Sector

Leveraging Continuous Process Improvement to Build Better Governments

Brandon Cole

ASQ Quality Press
Milwaukee, Wisconsin

American Society for Quality, Quality Press, Milwaukee, WI 53203
© 2011 by ASQ
All rights reserved. Published 2011.
Printed in the United States of America.

17   16   15   14   13   12   11   5   4   3   2   1

Library of Congress Cataloging-in-Publication Data

Cole, Brandon, 1982-
Lean-six sigma for the public sector: leveraging continuous process improvement to build better governments/Brandon Cole.
   p. cm.
Includes index.
ISBN 978-0-87389-806-5 (alk. paper)
1. Total quality management in government—United States. 2. Process control. I. Title.
JK468.T67C65 2011
352.3'4—dc22
                                    2011003214

No part of this book may be reproduced in any form or by any means, electronic, mechanical, photocopying, recording, or otherwise, without the prior written permission of the publisher.

Publisher: William A. Tony
Acquisitions Editor: Matt T. Meinholz
Project Editor: Paul O'Mara
Production Administrator: Randall Benson

ASQ Mission: The American Society for Quality advances individual, organizational, and community excellence worldwide through learning, quality improvement, and knowledge exchange.

Attention Bookstores, Wholesalers, Schools, and Corporations: ASQ Quality Press books, video, audio, and software are available at quantity discounts with bulk purchases for business, educational, or instructional use. For information, please contact ASQ Quality Press at 800-248-1946, or write to ASQ Quality Press, P.O. Box 3005, Milwaukee, WI 53201-3005.

To place orders or to request ASQ membership information, call 800-248-1946. Visit our Web site at www.asq.org/quality-press.

∞ Printed on acid-free paper

Quality Press
600 N. Plankinton Ave.
Milwaukee, WI 53203-2914
E-mail: authors@asq.org
The Global Voice of Quality™

Dedication

This book is dedicated to my loving wife Hillary, my supportive family, and our dog Milly. I would also like to thank everyone who provided insights, thoughts, and "war stories" for this book. Keep in mind that the work we do is a journey.

Contents

List of Figures and Tables ix

Preface xv

Introduction – The Forcing Function 1

Section 1: Overview of the Lean-Six Sigma Methodology 3
    Process Improvement 3
    Lean 3
    Six Sigma 4
    Integration of Lean and Six Sigma 5
    DMAIC 6
    Design for Six Sigma 8
    Value of Phase Tollgates 10

Section 2: Challenges of Using Lean-Six Sigma in the Public Sector ... 13
    Hierarchical Environment 15
    Limited Sense of Urgency 16
    Leadership Support for Enterprise Programs is Difficult to Obtain ... 18
    Not Profit Focused 21
    Lack of Common Goals 22
    Lack of Customer Focus 24
    High Employee Turnover 26
    Complexity of Public Sector 27
    Mix of Employee Types 29

Section 3: Laying the Groundwork for Success 31
Expectation Setting. 31
Key Project Stakeholders and Team Members 32
    Executive Leadership and Deployment Champions 32
    Process Owners 32
    LSS Master Black Belts 33
    LSS Black Belts 33
    LSS Green Belts 33
    LSS Yellow Belts. 33
    LSS White Belts 33
    Functional Subject Matter Experts 34
    Accounting and Finance. 34
Executive Steering Committee 34
Integrated Stakeholder and Communication Plan 36
Risk Management Plan 40
Strategic Planning. 42
SMART Goals and Objectives. 44
Visible Balanced Scorecards and Metrics 46
Create a Funding Re-allocation Plan 49
Tailored Approach 50
Where Do I Get My Projects? 51
    SWOT Analysis 52
    Project Selection Meeting 53
    Brainstorming. 55
    Lean-Six Sigma Audit Worksheets 56
    Affinity Diagrams 57
    Multi-voting 58
    PICK Diagram 58
Why Lean-Six Sigma Programs Fail. 61
Spotlight – Building Effective Teams. 63

Section 4: Focus on Waste, Then Variation. 65
Value of a Project Kickoff Meeting 65
Project Charter – One Stop Shop for Why the Project is
    Being Pursued 66
    Business Case 67
    High Level Description of Current Process 67
    SMART Goals and Objectives 67
    Goal vs. Objective Statement 68
    Quantitative vs. Qualitative. 68
    Project Scope. 68

High-Level Milestones	69
Example Project Charter	69
SIPOC	70
Vo"X"	72
Types	72
Customer Segmentation	74
Gathering Vo"X"	75
Critical-to-Quality (CTQ) Elements	79
Kano Analysis	81
Integrated Vo"X" Approach	83
Baseline Enterprise-wide Process Maps	84
Standard Operating Procedures	94
Spaghetti Diagrams	97
Kaizen	100
5S	102
Sort	103
Set in Order	104
Shine	104
Standardize	104
Sustain	105
Safety	105
5S Events – How do I Implement 5S?	105
Future State Process Mapping	107
Leverage Fast, Visible, Highly Impactful Successes	112
Spotlight – Facilitating Effective Meetings and Events	113
Section 5: Basic Quality Tools	**115**
Measurement Plan	116
Benchmarking	120
Run Charts	121
Fishbone Diagrams	123
Check Sheets	126
Control Charts	126
Histograms	132
Pareto Charts	135
Scatter Diagrams	138
Implementation Plans	141
Sustainment Plans	143
Spotlight – Pros and Cons in Utilizing Existing Data	144

Section 6: Create a "Buzz" **145**
    Stakeholder Involvement................................ 145
    Visibility 146
    Continued Communication is Critical...................... 146
    No Such Thing as a Problem, Just Opportunities 147
    Awards and Recognition in the Public Sector 147
    Spotlight – Importance of Soft Skills in Lean-Six Sigma...... 148

Section 7: Sustainment...................................... **149**
    Dealing with Turnover................................... 149
    Project and Data Transparency 150
    Civilian Component..................................... 151
    Importance of Good Governance......................... 151
    Capturing Lessons Learned.............................. 152
    Repository of Best Practices.............................. 152
    Repository of Future Opportunities....................... 153
    Agility... 154
    Using Lean-Six Sigma for Organizational Design........... 154
    Spotlight – Performance Measurement 155

Section 8: Conclusion....................................... **157**
    How Have We Overcome All the Public Sector Challenges?... 157
    There Will Be Iterations Across the Phases 158
    Other Process Improvement Methods..................... 158
    Even Change Leaders Need to Continuously Adapt 160
    Importance of Leadership Buy-in......................... 160
    Continuous Improvement............................... 162
    Spotlight – Greening Your Organization Using
        Lean-Six Sigma..................................... 162

Appendix A – Interview Questions........................... **165**

Index *167*

List of Figures and Tables

Table 1.1	The phases of DMAIC	7
Figure 1.1	DMAIC vs. DMADV	9
Table 2.1	Approaches, tools, and methods for overcoming a hierarchical environment	15
Table 2.2	Approaches, tools, and methods for overcoming a lack of urgency	17
Table 2.3	Approaches, tools, and methods for overcoming a lack of leadership support	18
Table 2.4	Sample executive steering committee agenda	19
Table 2.5	Approaches, tools, and methods for overcoming a lack of profit focus	21
Table 2.6	Approaches, tools, and methods for overcoming a lack of common goals	23
Table 2.7	Approaches, tools, and methods for increasing customer focus	25
Table 2.8	Approaches, tools, and methods for overcoming employee turnover	26
Table 2.9	Approaches, tools, and methods for overcoming complexity	28
Table 2.10	Approaches, tools, and methods for overcoming a mix of employee types	30
Table 3.1	Impact of the process/project on stakeholder	36
Table 3.2	Power to transform the process/project	37
Table 3.3	Commitment to process/project change	37
Table 3.4	Level of expertise on the process/project	37
Table 3.5	Communication strategy	38
Figure 3.1	Sample communication plan/stakeholder analysis	39
Table 3.6	Example risk management plan	41
Table 3.7	Example risk mitigation plan	41
Figure 3.2	Relationship between LSS strategy and execution	42
Table 3.8	Example mission and vision statements	43
Table 3.9	Example goals and objectives	44
Table 3.10	SMART goals	45
Figure 3.3	Theoretical balanced scorecard	47
Table 3.11	Example funding reallocation plan	49

Table 3.12	SWOT analysis.	52
Table 3.13	Example LSS audit worksheet.	56-57
Figure 3.4	Example affinity diagram	58
Table 3.14	PICK diagram categories.	60
Figure 3.5	Example PICK diagram template.	61
Table 4.1	Sample project kickoff agenda	66
Figure 4.1	In-and-out-of-scope tool.	69
Table 4.2	Example project charter	70
Figure 4.2	SIPOC template.	71
Table 4.3	Example of customer segmentation	74
Figure 4.3	Relationship of Vo"X" cost vs. depth of information	75
Table 4.4	Example interview agenda	76
Figure 4.4	A robust survey approach	77
Figure 4.5	Example CTQ tree.	80
Figure 4.6	Kano model	82
Table 4.5	Example Kano analysis output table	82
Figure 4.7	Integrated Vo"X" approach.	83
Figure 4.8	Enterprise-wide value stream.	85
Figure 4.9	Granularity of process maps required to achieve respective goals	86
Figure 4.10	Correlation of LSS initiatives, methods, and maturity	87
Table 4.6	EWPM checklist.	88
Figure 4.11	Example high-level process map	89
Figure 4.12	Example process step 5: make brownies (manufacturing)	91
Table 4.7	Types of waste.	93
Table 4.8	Value analysis criteria.	94
Table 4.9	Example SOP for process map 5: make brownies	96
Figure 4.13	Example spaghetti diagram.	98
Figure 4.14	Example spaghetti diagram improved	99
Figure 4.15	Sample kaizen agenda	101
Table 4.10	Fundamentals of 5S.	103
Table 4.11	Sample 5S agenda and timeline	106
Figure 4.16	Example current state process map	108
Figure 4.17	Example value stream analysis.	110
Figure 4.18	Three major aspects that impact every organization.	111
Figure 4.19	Example streamlined future state process.	111
Figure 5.1	Measurement plan linkages.	116
Table 5.1	Example discrete and continuous measurements	117
Figure 5.2	Call answer speed.	118
Table 5.2	Example measurement plan	119
Figure 5.3	Example run chart.	122
Figure 5.4	Five whys example.	124
Figure 5.5	Example fishbone diagram	125
Figure 5.6	Check sheet approach to data gathering	126
Figure 5.7	Attribute data control charts	127
Table 5.3	Attribute control limit calculations	128
Table 5.4	X-Bar and R control limit calculations	128
Table 5.5	X-Bar and R control chart constants.	129

Figure 5.8	Control chart zones		130
Figure 5.9	Example control chart		131
Figure 5.10	Example control chart with out-of-control indicators		132
Figure 5.11	Example normal distribution histogram		133
Figure 5.12	Example non-normal distributions		134
Figure 5.13	Additional non-normal distributions		134
Figure 5.14	Example Pareto chart		135
Figure 5.15	Pareto chart used to illustrate data stratification		137
Figure 5.16	Scatter diagram exhibiting a positive correlation		138
Figure 5.17	Scatter diagram exhibiting a negative correlation		139
Figure 5.18	Scatter diagram exhibiting a non-linear correlation		139
Figure 5.19	Correlation types		141
Figure 7.1	Robust governance structure		151
Figure 7.2	LSS for organizational design		155

Preface

The information, examples, and insights provided as part of this book do not represent the opinion of my current or past employers. The intent is to describe pitfalls I have encountered in the public sector and provide examples of how other Lean-Six Sigma (LSS) experts and I have overcome these obstacles (although they are masked to protect the identity of specific organizations and team members). I also want to provide specific tools for overcoming challenging situations. This book is not meant to demonstrate that the public sector is not willing and driving change, but rather to show that public sector leadership are willing to persevere, to overcome obstacles, and to implement true process improvement. Most of the leaders I have worked with in the public sector are driven to provide the support necessary, as part of their organizational charter, to deliver the highest quality output for their respective customers. Some organizations in the public sector are more mature than others when it comes to LSS. There are even pockets of excellence, but one item remains constant: a willingness to change, to drive improvements and efficiencies, and to become highly effective. I want to thank each and every organization I have interacted with for being open to the LSS methods and helping to drive an organization of true continuous process improvement.

Introduction – The Forcing Function

Six months = millions of dollars. This was the impact of a recent public sector LSS initiative. Results such as these are not out of the ordinary using the LSS toolkit, even in the public sector. Leaders everywhere are required to "do more with less," enhance budget and organizational performance, and identify innovative ways to increase their impact. One process improvement methodology that has demonstrated time and time again that it can assist in all of these areas is LSS. Senior leaders across all industries—from pharmaceutical, automobile, and semiconductor manufacturers to financial services, construction, and information technology organizations—have seen significant returns from implementing LSS. This can also be achieved in the public sector.

Similar to people starting a new hobby, most organizations want to dive into LSS and attempt to apply all the tools in the toolkit without first developing a solid understanding of the pitfalls associated with implementing LSS in the public sector environment. One result of this can be "paralysis by analysis." It's possible to collect so much data that the information becomes overwhelming. The purpose of this book is to discuss the challenges of implementing LSS in the public sector and offer a tailored approach to overcome these obstacles.

Keep in mind that LSS is continuous process improvement, a pursuit of perfection within your organization. There is time to slowly move into the more advanced toolkit (for example, design of experiments, multi-vari charts, and regression analysis), but we will focus initially on institutionalizing the enterprise-wide approach, obtaining the voice of your customer, and creating an environment with appropriate performance indicators to measure success. Although the advanced concepts can provide significant value, using even the most basic tools, paired with a systematic approach, can result in immediate impacts to your organization.

The LSS methodology is not meant to be radical, "big-bang" change. LSS focuses on eliminating waste (lean) and then reducing variation (Six Sigma) in a process or product that contains no waste. There is no benefit to eliminating variation in a broken process or product. That would just make the defects or defective outputs more robust.

The information covered in this book will allow someone to have an immediate impact in any public sector organization. Organizations continuously try to make themselves better by reducing cost, increasing customer satisfaction, and creating an environment of empowered employees who continuously strive for excellence in each process and product. This book describes some of the most powerful continuous process improvement tools and also provides insights in the form of spotlights, key takeaways, and potential pitfalls. These lessons learned can be invaluable in assuring that your use of the LSS toolkit is successful.

Beyond the quality tools included in the toolkit, some areas are often overlooked as projects are being completed. One of the most critical aspects is a dedicated focus on change management and communication. People are typically averse to change, especially when they do not understand why the change is occurring, when they did not know that it was coming, and when they were not involved in the solution. To address that reluctance, change management tools and techniques are integrated throughout this book. Even if you have identified a solution that could save the organization millions of dollars, you must also communicate the *who, what, when, where, how,* and, most important, the *why*. If you skip this step, the potentially most valuable change is not likely to be successfully implemented or sustained.

The intent of this book is to not discuss every LSS tool that exists or describe the most advanced tools in excruciating detail (entire books are written about some of the tools covered here). Rather, I will focus on the tools I believe will help improve a public sector organization without requiring extensive amounts of training or ramp-up. Training is definitely crucial as part of a long-term LSS program, but I want you to be able to make an impact *today*. Once leaders have seen the value of LSS and you have gained some momentum, you can work to launch an enterprise-wide LSS program.

Because the public sector spans such a diverse range of organizational charters (such as transportation, education, and defense), this book will not focus solely on either manufacturing or services. Rather, it will attempt to provide a balanced approach to utilizing LSS in all environments. We will provide specific examples of how tools can be applied and what the output means from both a services and manufacturing perspective. I hope you are excited about the journey because it will be a fulfilling ride, especially if you follow the guidance provided in this book.

Section 1
Overview of the Lean-Six Sigma Methodology

"To improve is to change; to be perfect is to change often."
– WINSTON CHURCHILL

PROCESS IMPROVEMENT

Process improvement has existed for hundreds of years, in the creation of the first printing press, which set the stage for mass production, and later in the development of Henry Ford's modern-day assembly lines and factories. Improving an organization is a continuous journey, rather than a big bang, once-and-done transformation. Improvement methods have changed over time, but the fundamental pursuit of making products, processes, and services better remains central to the success of today's organizations.

The difficult part of any process improvement is instilling it as part of the DNA of an organization and making it part of everyone's job function, not just a responsibility of the LSS program office or another specific support function. This can be accomplished by ensuring that people involved in LSS projects are empowered to identify areas for improvement and made to feel as though they can truly make an impact on their job roles. If those three items are accomplished, which is no easy task, resources will be positively driven to improve the organization.

LEAN

In simple terms, lean is the elimination of waste within a process or product. In this definition, waste is anything that does not directly add value in the customer's eyes. This improvement approach focuses on the entire value-stream, from a raw materials supplier perspective to the ultimate customer and even looping back into the stream in the form of post-customer reuse or recycling where possible. By focusing on the customer and the value stream, you can improve your process as a secondary effect. For example, reduced costs can result in reduced costs to the customer. An increase in quality may result in increased customer sales. The focus of lean is to deliver only what the customer wants and needs (that is, waste-free products and services).

The toolkit for lean includes common sense improvement tools that can help you focus on high-impact, quick wins. This can help you gain significant leadership buy-in early in your deployment by providing an approach that is easy to understand and that results in significant, quick, sustainable results.

Used alone, lean can result in important improvements, but there is one drawback. Even if you eliminate all the wastes in a process, you may still have significant unseen variation if you lack statistical tools to identify variation in your process, product, or service.

SIX SIGMA

Six Sigma, a term first coined by Motorola, focuses on the elimination of variation. In statistical or theoretical terms, Six Sigma is a process that is able to limit defects to only 3.4 per million opportunities. From a reality or continuous process improvement methodology stand-point, Six Sigma is a collection of management and statistical tools that are used to identify and sustain high-value improvement efforts.

There is one significant pitfall to Six Sigma. It focuses only on the reduction of variation within an existing process and does not address whether all the steps in the process add value. As an example, consider the variation in making pizza. Six Sigma focuses only on ensuring that each large pepperoni pizza contains twenty pieces of pepperoni. It does not consider whether the customer actually wanted twenty slices of pepperoni. No attention is paid to overall customer value. This is what has led to the integration of the two methodologies, lean and Six Sigma.

> **Public Sector Project Highlight**
>
> **Not Every Process Should be Six Sigma**
>
> As you integrate the LSS methodology into your organization, remember that it's not necessary that every process reach Six Sigma levels (3.4 defects per million opportunities). The cost of some Six Sigma initiatives can be high with little return on investment. Some processes, products, and services are not mission critical. For example, it might not be necessary to optimize the weight of an ice cube to Six Sigma levels. If each proposed initiative is linked to your strategic plan and goes through a robust project selection process (we will discuss this later), project selection should not be an issue. Remember, there *is* such a thing as diminishing returns on quality. Six Sigma levels are essential on such things as airplane parts and medical equipment. Decisions must be based on sound analysis.

INTEGRATION OF LEAN AND SIX SIGMA

The integration of the lean and Six Sigma methodologies creates a "best of both worlds" approach. It is the fundamental integration of an improvement toolkit that focuses on the elimination of waste *and* the elimination of variation. In terms of displaying immediate value in your organization, focus on areas where mass amounts of waste can be eliminated quickly. This is known as "kaikaku." Then focus on lean. Address these areas first and achieve success in order to gain leadership and stakeholder support for your overarching LSS efforts. This will assist in gaining critical leadership buy-in, create a collaborative environment across organizational stovepipes, and demonstrate that the focus is not on eliminating specific positions or resource functions but rather on creating a streamlined public sector organization. Once you have gained momentum and begun to reduce the amount of "low-hanging fruit," focus your efforts on reducing variation in the streamlined process through Six Sigma.

The integration of the two improvement methods helps focus initiatives on areas such as cost reduction, work-in-process inventory, and space requirements while increasing productivity, quality, and overall customer satisfaction. These are the most important aspects to sustaining your organization's relevance. The goal is to use a systematic approach to problem solving in order to foster and sustain an environment of continuous process improvement. Working toward 3.4 defects per million opportunities may sound like a difficult and complex project, but don't let that defeat your enthusiasm. This book will provide insights into typical roadblocks and explain methods you can use to overcome them using LSS tools to identify, implement, and sustain high-value initiatives.

We will discuss the basic LSS tools, and you will use some of them on each project you take on. It's important that you use only the tools necessary to achieve your goal statement as created in the project charter (more to come on that topic later). Before we introduce the tools of LSS, we first must understand some important terminology.

As a newer practitioner, it is crucial that you remember that appropriately marketing the value of LSS on the front end can help ensure a lasting transformation. Your job is to help people feel excited about what is coming and keep them excited until the results are delivered and sustained. It's important to obtain buy-in from all parties (leadership, functional experts, and other project participants). This is critical to the deployment, project success, and sustainment of your LSS program.

A pool of personnel will want to be involved including Master Black Belts, Black Belts, Yellow Belts, Project Team Members, Sponsors, and Champions. These resources can be internal or external to your organization as long as they willingly support the LSS work. Identify the ones who are not supportive and create an appropriate communication and risk mitigation plan. The members of the organization will be providing data, input, and overall support. *If you are not involving them throughout the process, they will not be supportive of the changes.*

DMAIC

LSS DMAIC is a structured, systematic, and phased approach to solving problems in an existing process, product, or service. It is a process improvement framework focused on maximizing both efficiency and effectiveness. DMAIC, which stands for Define, Measure, Analyze, Improve, and Control, is the systematic analytical approach that is the underlying concept of most LSS projects (there is also the Design for Six Sigma methodology, which will be discussed later). Table 1.1 is a high-level representation of what is involved in each of the DMAIC phases.

Although LSS is a systematic approach, this does not mean that the various DMAIC phases are not iterative in some respects. For example, you may discover during the Analyze phase of your project that you missed a necessary data point in the Measure phase. Or, you may discover during the Improve phase that you missed a key stakeholder in the Define phase. In situations such as these, do not hesitate to complete the necessary steps to ensure a successful project. If a positive outcome requires revisiting a previous phase for additional information, include this information in your next project update with leadership.

Table 1.1 The phases of DMAIC.

Define
• Accurately define and scope the project
• Identify key performance improvement metrics
• Baseline current process at a high level
• Identify stakeholders
• Gain leadership buy-in for the project
Measure
• Identify data to support the project
• Create a plan for gathering data
• Identify and verify the measurement system
Analyze
• Identify methods to analyze the data
• Identify root causes tied to improvement metrics
• Determine "true" causes for identified root causes
Improve
• Identify improvements for "true" causes
• Quantify the impact of improvements on performance metrics
• Gain leadership buy-in for implementation
Control
• Create a method to sustain improvements
• Foster continuous process improvement
• Gather lessons learned
• Celebrate success
• Submit quantifiable improvements for finance validation

The first thing that every project leader should do is accurately define the problem to be improved. The basic tools used in this phase are a project charter, stakeholder analysis, communication plan, project plan, SIPOC, and a high-level process map. In addition to these basic tools that are normally used in a LSS project, other more advanced tools are also included within the LSS toolkit.

The next phase in the DMAIC process is the Measure phase. This is the phase in which we identify data that will help to support our conclusions. The ultimate goal of any LSS effort is to use fact-based decisions to make an assessment of areas in which improvements are required. The tools normally used in this phase include a measurement plan, detailed process mapping, value-stream mapping, and the voice of the customers, employees, administration, and suppliers who support the process or product. Now that we have data to support conclusions, we must analyze this data to identify trends or other areas for improvement.

Tools that can yield significant results in this area include Pareto charts, scatter diagrams, and cause-and-effect analysis. Once you have reached fact-based conclusions on possible opportunities for improvement, you can use improvement approaches such as kaizen, 5S, and brainstorming tools to create improvement and implementation plans.

The next step is to implement the actual improvements once they have been signed off on by senior leadership and all other appropriate key stakeholders. This ends the process, right?

No, this is where the most crucial step occurs, *control* or sustainment. In the Control phase of the project you will use tools such as control charts and standard operating procedures and capture lessons learned and celebrate the project successes. The high-level outline of the DMAIC process demonstrates the value in the approach; it is structured and systematic.

The keys to a robust LSS approach include:

- Following the phased structure of the methodology ensuring continued leadership buy-in throughout each phase,
- Creating a fundamental coaching, mentoring, and training structure for new LSS belts, and
- Fostering an open and collaborative environment of continuous process improvement where employees feel as though they have a "voice" and can openly share ideas for improvement (in a non-attribute manner) that will change their organization.

DESIGN FOR SIX SIGMA

There are two major phased approaches in the LSS toolkit. The first is DMAIC, which we have already discussed. The second is DMADV (also known as Design for Six Sigma or DFSS). It's beneficial to understand this at a high level even if it is not used by the organization. DMADV (which stands for Define, Measure, Analyze, Design, and Validate) focuses on building a process, product, or service utilizing a LSS approach. That is, building the product, process, or service with significant consideration of the value provided to the customer and in creating a robust and repeatable process that limits the possibility of variation. A simple approach to determining when to use DMADV and when to use DMAIC is outlined in Figure 1.1.

Keep this in mind: although you may start out an initiative to improve an existing process using the DMAIC approach, you may find during the Improve phase that a new process, product, or service is required. In a case such as this, you could transition to the DMADV approach for that specific project. With the increased focus on building to customer specifications, you might find that you need some of the critical tools involved in a DMADV effort: voice of the customer, supplier, partner, or employee (Vo"X"), quantified critical to quality elements (CTQ), quality function deployments (QFD), failure mode and effects analysis (FMEA), and other tools that focus on integrating customer value into a new product, process, or service. If it is determined that DMADV is the appropriate approach, please refer to another text outlining this methodology in more detail.

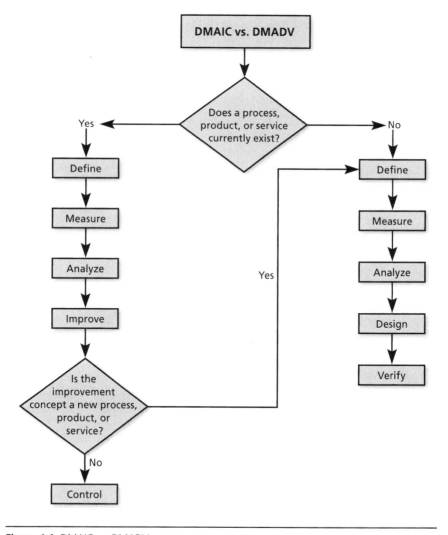

Figure 1.1 DMAIC vs. DMADV.

VALUE OF PHASE TOLLGATES

An important part of the DMAIC process is the concept of phase tollgates. These are milestones that allow the team to review and approve progress points. Can you skip the tollgates? The answer to this question is…it depends. It is invaluable to go through the process of an internal tollgate review with the project team, but it will ultimately depend on the time availability of senior leaders. If they are averse to continuous involvement you must, at a minimum, require them to be a part of the Define, Analyze, and Improve phase tollgate reviews. This is where they will review the identified problem areas and sign off on what you plan to implement, allowing you to go forward with the project as defined.

So what should be included in a tollgate review? The most positive tollgate reviews focus on being concise. I recommend reviewing the following items with senior leaders at each phase:

- **Define Phase** – Project charter sign-off, stakeholder analysis, and SIPOC review. These tools will allow leaders to review the opportunity and goal statements and will provide further details on stakeholder involvement and the scope of your overall initiative.

- **Measure Phase** – Measurement plan, current state process map, and output from any Vo"X" analysis. These tools will provide information on resources required to capture valuable data, more detail on the overall process, and any information gained from the customers, suppliers, and employees.

- **Analyze Phase** – Any information gleaned from scatter diagrams, Pareto charts, cause-and-effect analysis, and other basic quality tools. This information will provide initial insights on opportunities for improvement.

- **Improve Phase** – Kaizen, 5S, and brainstorming results, as well as improvement and implementation plans. These will provide not only the recommended improvements for your overall project, but also the proposed implementation plan (including stakeholder, resource, and time requirements).

- **Control Phase** – Standard operating procedures, control charts, and plans for communicating and celebrating the success of your initiative. It is important to support an environment of continuous improvement while also ensuring improvements implemented are sustained and celebrated.

Each tollgate review will be different based on which tools were used in each phase and the conclusions reached. Overall, the goal of each tollgate review is to gain the necessary resources and sign-off that will allow you to move on to the next phase of the initiative. This way, when you reach the Improve and Control phases, leaders will already be supportive of most or all of the improvements.

Public Sector Project Highlight

Tollgates Are More than Hard Data

Tollgates provide more value than just an opportunity to review the project data with senior leadership. A tollgate review is also an opportunity to support the overall change management strategy. In advance of the tollgate meeting, especially if it is one of the first ones in an organization, you can add significant value by briefing senior leaders on potential questions they may ask, their role in breaking down barriers for the LSS team, and the need for them to instill further excitement, confidence, and enthusiasm for the project and the LSS program overall. If the LSS team leaves the room with a positive feeling toward the work completed, this will further engage the current team and also permeate across the organization. As a result, people will begin to view the LSS program positively and want to be a part of the improvements.

Section 2

Challenges of Using Lean-Six Sigma in the Public Sector

> *"Obstacles are those frightful things you see when you take your eyes off your goal."*
>
> – HENRY FORD

The LSS tools can be applied in nearly every environment. Government and public sector organizations are no exception. This does not mean that there are not certain considerations that must be taken into account prior to using the approach. The primary challenges I have seen, faced, and overcome in the public sector include the lack of profit focus; a lack of customer focus; a lack of common goals across the enterprise; an inconsistent or complete lack of leadership support; a constraining hierarchical environment; limited or lack of urgency; an increase in complexity of organizational interactions and missions; a mix of civilian, military, contractor, union, and non-union personnel; and, finally, significant and consistent turnover across all levels of public sector organizations.

As part of this section, I want to share insights from a varied sample of senior LSS practitioners who have experience in both the public and the private sectors. The people I interviewed have an average of 25 years of total work experience (a few exceed the 40-year mark) and approximately 16.5 years of LSS related experience. This experience was gained in a variety of industries including aerospace, automotive, academia, chemicals, semiconductors, power systems, government, information technology, consulting, healthcare, and consumer products. Their path to gaining LSS certification was also varied. Some actually launched the initial process improvement initiatives within their organization. Others used earlier methods such as TQM. Most progressed up through the normal LSS ranks of Green Belt, Black Belt, and finally Master Black Belt. Now they are all focused on LSS deployment, implementation, and sustainment in the public sector, bringing years of successful LSS experience to overcome the complexity of government organizations.

One thing was clearly similar in all of the interviews: the public sector is different from the private sector when it comes to deploying, implementing, and sustaining LSS. Integrated throughout this book are the thoughts, lessons learned, and input from these senior practitioners about how the sectors are different and about methods and approaches to overcome the challenges. (For a full list of the questions asked as part of these interviews, please see Appendix A.)

As part of each interview, participants were asked to rank the nine public sector challenges included in this section. The result of this question was the rationale for the order of each discussion item and the outcome was as follows (ranked from most challenging to least challenging):

1. Hierarchical or stovepiped environment
2. Limited sense of urgency
3. Lack of leadership support
4. Lack of profit or revenue focus
5. Lack of common goals
6. Lack of customer focus
7. High employee turnover
8. Complexity of the public sector
9. Mix of various employee types

We also discussed which tool they believe is the most powerful, what they believe the future holds for LSS, how perceptions in the public sector are changing relative to the methodology, and what benefits they have achieved using the approach to build better governments.

The results of these interviews, as well as my personal experiences implementing LSS in the public sector, are discussed throughout this book. As part of every LSS journey, continued value is provided by sharing best practices, understanding lessons learned, and hearing how the approach has been so successful in some areas of the public sector. This information has been included throughout the text and highlighted in more than thirty Public Sector Spotlights (which discuss lessons learned, tips, traps, best practices, and results). I have also provided a simplified table to show the multitude of tools that can be used to overcome each of the challenges (see Table 2.1 – Table 2.3 and Table 2.4 – Table 2.10). The data demonstrates that perseverance, leadership support from middle managers and senior leaders, and the creation of a culture of proactive change can lead to continuous process improvement.

HIERARCHICAL ENVIRONMENT

Table 2.1 Approaches, tools, and methods for overcoming a hierarchical environment.

• Executive steering committee • Integrated stakeholder and communication plan • Strategic planning • Funding reallocation plan • SWOT analysis • Project selection • Brainstorming	• Lean-Six Sigma audit • Multi-voting • PICK diagram • Project kickoff meeting • Project charter • SIPOC • CTQs • Run charts	• EWPM • Kaizen • 5S • Future state mapping • Measurement plan • Implementation plan • Sustainment plan

The number-one difference identified between the private and public sectors is a rigid, hierarchical environment, sometimes called a stovepiped environment. Feedback from the interviewees provided evidence that this area affects every aspect of a LSS program (deployment, implementation, and sustainment). The stovepiped environment was responsible for "islands of excellence," where the LSS program had significant momentum and results achieved in a relatively small area of the organization. It also caused failures in areas that had previously seen no results because of a lack of willingness to support change.

As an example of the issues surrounding a stovepiped environment, one interviewee described the process to complete a recent kaizen event outbrief. The project leaders were required to brief the proposed improvement, implementation, and sustainment plans up three levels in their respective organizations. Unfortunately, the improvements identified also affected three different organizations across the enterprise. In order to implement all the changes, each project leader, along with his or her leadership chain, had to visit three other project leaders to gain support to brief the next three levels down in their respective organizations. Once all was said and done, project leaders had to brief the improvement, implementation, and sustainment plans 12 different times before they were able to move forward with the first step in their action plans (approximately 100 days after the completion of the actual improvement event).

Fortunately, in this example, the project leader ultimately was able to gain support from all the necessary stakeholders. In some instances this cannot be achieved and the improvement team may need to reconvene to identify alternative improvement solutions. Supportive champions can work to break down these stovepipe roadblocks. If not removed, they can be detrimental to the long-term success of any LSS program.

16 Section Two

> **Public Sector Project Highlight**
>
> **Stovepipes**
>
> As part of a LSS project involving classified data, stovepipes were a requirement due to the potential security risks of sharing information across the enterprise. This required the public sector agency to communicate the value stream using disparate process maps. The agency overcame the possible sub-optimization of the process by having an outside stakeholder, who was cleared to view all of the process maps (minus the classified data, of course), provide an integrated look, which resulted in significant efficiency gains. This was an example of how challenges can be overcome by communicating the potential risks and obtaining perspective from all stakeholders to identify potential mitigation strategies.

LIMITED SENSE OF URGENCY

An apparent lack of urgency was the second most pervasive difference between the private and public sectors. This was often evidenced by two common themes: a significant amount of continued turnover in the public sector, where resources were able to freely move across the enterprise (a high risk when identifying potential belts, champions, and sponsors), and a lack of communication relative to "What's in it for me? (WIIFM)."

The first roadblock, dealing with turnover, will be discussed in further detail as part of Section 7. Clearly articulating WIIFM is something that can be easily accomplished through consistent communication. It must be clearly stated as part of the initial training through messages about career development, the ability to identify improvement opportunities, and making the organization a better place to work.

As part of each improvement initiative, the impact of the project must be clearly integrated as part of the stakeholder communication plan and describe the impact on individuals and the organization as a whole. Finally, as part of each project outbrief and sustainment plan, the impact of efficiencies or effectiveness gains should be clearly communicated. This helps ensure that the LSS program communicates WIIFM at every opportunity.

Another approach to overcoming a limited sense of urgency is to provide increased accountability (see Table 2.2). The ownership of action plans was a continued theme in terms of an opportunity for improvement in the public sector. This can be achieved through the establishment of a robust governance structure with an executive steering committee that focuses on the completion of action plans and sustainable results and on the integration of action items as part of mid-level status meetings that require individuals to identify LSS action items, their respective status, the anticipated date of closeout, and potential roadblocks to completion. The identification of roadblocks helps individuals know that leaders are willing to provide appropriate support if a roadblock presents itself.

Table 2.2 Approaches, tools, and methods for overcoming a lack of urgency.

• Executive steering committee	• Project kickoff meeting	• Benchmarking
• Risk management plan	• Project charter	• Fishbone diagrams
• Strategic planning	• Vo"X"	• Control charts
• Visible balanced scorecard	• CTQs	• Histograms
• Funding reallocation plan	• EWPM	• Pareto charts
• SWOT analysis	• Kaizen	• Scatter diagrams
• Project selection	• 5S	• Implementation plan
• Lean-Six Sigma audit	• Future state mapping	• Sustainment plan
• PICK diagram	• Measurement plan	• Run charts

Public Sector Project Highlight

Using Urgency to Overcome Roadblocks

When completing a recent LSS initiative, the LSS MBB (Master Black Belt) realized that in order to implement the proposed system modifications an investment of less than $10,000 would be required. Instead of requesting the funding as part of the implementation plan, the LSS MBB decided to take a more proactive approach. This involved completing a cost-benefit analysis that clearly showed the linkages between the $10,000 investment and the anticipated cost efficiencies (approximately $50,000 in cost savings annually). As part of the cost-benefit analysis, the LSS MBB also addressed the impact on customer morale and the speed with which efficiencies could be gained. This turned a possible difficult decision into a "no-brainer" by clearly showing how quickly the investment could be recovered and the future efficiency and effectiveness gains (that is, by creating a sense of urgency by clearly articulating how quickly the savings could be gained).

LEADERSHIP SUPPORT FOR ENTERPRISE PROGRAMS IS DIFFICULT TO OBTAIN

The chief executive officer (CEO) is the standard executive support expected as part of a successful LSS program in the private sector. This support is gained through executive training and by clearly articulating the impact LSS can have on the organization's future (for example, increased customer satisfaction, revenue, and/or profits). The positive impacts of the LSS program will generally directly impact the success and metrics of the CEO (for example, financial incentives and long-term career vitality). Therefore, the leaders are incentivized to ensure that everyone in the organization supports the program and makes LSS projects or training part of a career development plan.

In the public sector environment the rules and metrics are slightly different and the approach to gaining leadership support must be slightly tailored. LSS programs are often viewed as a one-time checkbox in this space, meaning that once employees have completed initial training and the required number of projects for certification, they go back to their primary job responsibilities and are not expected to maintain involvement in the program. This makes it difficult to sustain LSS efforts because principles are not integrated into their "day jobs" and are treated as more of a secondary responsibility.

From a leadership perspective, there are sound approaches for overcoming these differences: executive training, funding reallocation plans, governance, and executive steering committees. (See Table 2.3.)

Table 2.3 Approaches, tools, and methods for overcoming a lack of leadership support.

• Executive steering committee • Tailored executive training • Integrated stakeholder and communication plan • Strategic planning • Visible balanced scorecard • Funding reallocation plan	• SWOT analysis • Project selection • Lean-Six Sigma audit • PICK diagram • Project kickoff meeting • Project charter	• EWPM • Kaizen • 5S • Future state mapping • Implementation plan • Sustainment plan

The training for public sector leadership should be tailored specifically to each organization. It should clearly define targeted goals of the LSS program (for example, efficiency and effectiveness measures) and how any savings will be used (through a funding reallocation plan). This training should include the roles and responsibilities expected of leadership, such as signing off on project charters, ensuring adequate resources are available to complete each project, and acting as a spokesperson for the LSS efforts. The training must also encompass customized simulations to demonstrate the value LSS tools can provide to a sustainable enterprise (for example, waste elimination).

Once leaders understand the concepts and the value LSS can provide to the organization, they must provide input and signoff to the overall LSS strategy. This includes the governance structure (that is, the executive steering committee and individual roles and responsibilities), as well as the integration of performance measures/targets for each sub-organization. The executive steering committee should meet at least every month and should attract membership from across the enterprise. This helps drive accountability for each and every project. See Table 2.4 for a sample agenda.

Table 2.4 Sample executive steering committee agenda.

1. Training review
   A. Number of candidates trained
   B. Upcoming training
   C. Training candidates
   D. Training issues
2. Project results
   A. Recent initiatives completed
   B. Initiatives in progress
   C. Troubled projects and resources required
3. Change management
   A. Communication of project successes
   B. Sharing of lessons learned and best practices
   C. Rewarding outstanding leadership or project achievement
4. Sustainment
   A. High-level strategic plan and scorecard review
   B. Belts redeployed on other initiatives
   C. Review of project pipeline

The executive steering committee must be involved at the front end of each project to ensure that people, time, funding, and other necessary resources are provided in order to prevent project failure. This includes subject matter experts who can talk appropriately to the process, product, or service and have access to decision makers when appropriate. The committee must demonstrate continued support by removing roadblocks and always vocalizing how the project will benefit the organization. At least one senior LSS practitioner should be part of this committee (for example, a LSS Master Black Belt). This person will provide immediate feedback and recommendations for any questions or concerns. The committee should also include leaders from each functional area within the enterprise (for example, logistics, financial management, contracting, and mission support). The senior LSS practitioner can also help support the scoping and developing of initial project charters relative to any improvement opportunities discussed as part of the meetings.

Leadership support and continued commitment are instrumental to the success of any LSS program. Other stakeholders will not support the initiatives if they believe management is unsupportive. Why would they, if participation will not gain them visibility, promotions, or good graces in the eyes of their leadership.

Public Sector Project Highlight

Leadership

During the various interviews, a common theme surrounding leadership began to surface. Although the CEO or similar role is clearly the leadership support level required as part of private industry, in the public sector it seems as though middle managers provide the most influence. The evidence provided focused on gaining and sustaining stakeholder support and accomplishing problem resolution. The middle managers were able to create excitement for each initiative both at the leadership level and lower levels because they had an intimate understanding of the results of each project. This allowed them to confidently brief the savings and provide feedback on how they used the cost savings. The "buzz" created at the middle levels of the organization also helped support enterprise-wide buy-in. Therefore, it is critical that leadership buy-in is not only gained at the senior leadership levels, but also at the middle manager or division levels.

Management support is critical to success, but if you are unable to achieve it at the beginning of your deployment, don't despair. This just means that you must focus on proving the value of LSS. This can be easily accomplished by identifying "low-hanging fruit" to demonstrate the speed and impact with which LSS can affect your organization. You may also want to focus on the utilization of tools that are easier to understand (for example, fishbone diagrams, value stream mapping, and Pareto charts). Another common concern is the terminology (for example, DMAIC, DFSS, kaizen). If this problem occurs, tailor it to your leadership. For example, some organizations in the public sector seem to prefer rapid improvement events over kaizen events. It is *using* the tools that provides the value, not their names. The goal is to obtain sustainable improvement results; it's not necessary to be an idealist to the methodology. Once you demonstrate a few quick successes, leadership will begin to be supportive. Organizations that have seen significant returns from their programs have gone all the way to integrate LSS into their culture. For example, some integrate LSS into career development plans: mandating it as a requirement for promotions, showing organization-wide metrics relative to the results gained, tying the program to incentives where possible, and deploying an enterprise-wide LSS dashboard with drill-down capability to individual departments being briefed to the executive steering committee in the organization.

NOT PROFIT FOCUSED

Surprisingly, the lack of profit focus was not the primary hurdle faced by the various experts interviewed. I had initially ranked this as the number-one difference between the private and public sectors, but data doesn't lie. (See Table 2.5 for a list of approaches, tools, and methods for overcoming a lack of profit focus.)

Table 2.5 Approaches, tools, and methods for overcoming a lack of profit focus.

• Executive steering committee • Strategic planning • Visible balanced scorecard • Funding reallocation plan • Project selection • Lean-Six Sigma audit • Project charter	• Vo"X" • CTQs • EWPM • Future state mapping • Measurement plan • Benchmarking	• Check sheets • Control charts • Histograms • Pareto charts • Scatter diagrams • Sustainment plan

In private industry, LSS has a foundational impact on the bottom line by increasing revenue, increasing profits by driving down costs, or increasing some aspect of customer satisfaction. The clear linkage to these efficiency and effectiveness measures is what has made the methodology powerful in almost every private industry. The public sector is quite different in that executing more effectively or efficiently may actually drive down the power of the organization. In the public sector, the size of a budget or the number of human resources involved often translates into "a seat at the table." A larger budget means there is more focus on a value proposition at that given time. For example, increases in the Department of Transportation budget typically mean that there is more focus on transportation issues; increases in the Department of Defense budget typically means that there is more focus on defense issues.

Each and every LSS project must overcome the roadblock of seemingly decreasing budgets and resources by refocusing those efficiencies on mission-critical items through a funding reallocation plan.

Another obstacle to using LSS in the public sector is the inability to easily obtain data to support projects. In the private sector, the metrics required are part of the normal operating framework (look at any balance sheet, 10K, or other industry financial report and you will find a variety of useful data), whereas in the public sector the metrics must be defined and appropriate data sources must be identified. This can be overcome in the early stages by using more lean tools. These tools typically focus on gathering the data as part of improvement events such as process mapping, 5S, or kaizen. They use a "go and see" or "walk the process" type of approach and provide quick, sustainable improvements focused on eliminating waste. This supplies the rationale to facilitate more cost- and time-intensive data gathering assessments and collection efforts as the LSS program becomes more mature.

Public Sector Project Highlight

Project Selection Meeting

As part of a recent project selection effort, one interviewee created a customized public sector approach. This approach integrated the normal aspects of a project selection meeting (opportunity generation, scoping, problem/goal definition, and stakeholder identification), with clear linkages to a funding reallocation plan. The outcome of this approach was not only the identification of mission-critical requirements to re-infuse cost savings, but also future buy-in for the benefits of the LSS efforts. The list of future initiatives, along with their respective funding reallocation plans, should then be posted in a visible location to gain continued excitement for the LSS program.

LACK OF COMMON GOALS

In a private sector company, goals are obvious (revenue and profit). In a public sector organization, it may be that disparate goals exist across the entire government enterprise (with no common linkage of strategic plans, goals, and objectives) and also within individual organizations and divisions. This disparity between goals can occur even within a smaller government agency. The inability to clearly link to a common goal in the public sector is often compounded by the lack of critical data. (See Table 2.6 for a list of approaches, tools, and methods for overcoming a lack of common goals.)

Table 2.6 Approaches, tools, and methods for overcoming a lack of common goals.

• Executive steering committee	• Lean-Six Sigma audit	• Kaizen
• Integrated stakeholder and communication plan	• Affinity diagram	• 5S
	• PICK diagram	• Future state mapping
• Strategic planning	• Project kickoff meeting	• Measurement plan
• Visible balanced scorecard	• Project charter	• Benchmarking
• Funding reallocation plan	• Vo"X"	• Check sheets
• SWOT analysis	• CTQs	• Implementation plan
• Project selection	• EWPM	• Sustainment plan
• Brainstorming	• Run charts	

So why are common goals or linkages to common goals so important to an organization and the success of a LSS program? Common goals are the "beat of the organizational drum" providing an overarching objective for every task. This can help build momentum and sustain excitement. It can also clearly indicate which process improvement initiatives provide value to the organization and which are simply "certification projects" (projects that don't affect the strategic goals and objectives of an organization, that are completed only to meet a certification requirement).

As an example, a common goal can be to provide access to public transportation for a specific urban area. This becomes the basis for every organization that has a role in supporting that goal. Will the project and/or process help increase access to public transportation for that area? If not, and if there are no linkages to other strategic goals and objectives, you must investigate the possibility of eliminating that process or project. Some of the most common "ground-fruit improvements" I have witnessed in an organization have to do with processes being completed on an ongoing basis simply because "that is how things have always been done." It is the responsibility of the LSS expert, with support from the organization, to identify these outdated legacy processes, eliminate them, and re-direct the funding and resources to more mission-critical initiatives.

Without the clear identification of common goals from the highest to the lowest levels of an organization (see strategic planning), it will be difficult to sustain a robust LSS program. A commonality in goals can be used to support all aspects of the organization, from congressional budget requests to aligning every resource with a mission-critical asset. It is the role of the organization to define these goals and consistently revisit them to ensure they are current and provide the framework for achieving their overall mission and strategic vision.

> **Public Sector Project Highlight**
>
> **Common Goals**
>
> As part of a recent LSS project, the inability to make enterprise-wide process changes was identified as a significant roadblock. Process changes that were recommended at the lowest level of an organization had to obtain approval for implementation enterprise wide; enterprise-wide changes recommend by headquarters were often not implemented at the lowest level due to the decentralization of budgets and management structures. To overcome this circular problem with gaining enterprise buy-in, the interviewee often had to facilitate multiple improvement events at each location and provide a consolidated improvement strategy back to headquarters. In order to gain support across the enterprise, the project leader then had to create a change management and communication strategy to gain buy-in from each organization across the enterprise. Although this required significant time and effort, the results ultimately provided for millions of dollars in savings each year. If there had been an executive steering committee in place at the start of this initiative, with representation from all areas of the enterprise, the requirement for all of the non-value-added meetings might have been avoided.

LACK OF CUSTOMER FOCUS

From a private sector standpoint, customers are typically clearly identified and defined. Customer service objectives exist for quality, price, quantity, and other performance metrics. Customers who are not satisfied have the ability to seek another supplier and provide feedback to other customers. All of this makes customers the true "voice" of business success in the private sector.

On the other hand, the public sector varies significantly. There is typically a lack of definition relative to what meets the needs of customers. Often customers are not even identified. To begin to understand what your true value proposition entails, you must involve external stakeholders as part of each public sector initiative.

> **Public Sector Project Highlight**
>
> **The Value of Customer Feedback**
>
> A recent LSS initiative was focused on employees leaving public service. As part of this initiative, significant defects were uncovered relative to paperwork associated with financial benefits such as accrued vacation days. The outcome of this effort was the creation of a standardized employee exit checklist (similar to a standard operating procedure) with examples and specifications provided for each process step. This created a simplified, self-service approach to completing the necessary requirements. This resulted in a decrease in defects and an increase in customer satisfaction (both primary goals for this initiative), as well as reduced quality assurance and defect remediation required for non-compliant customers (both secondary goals for this initiative). This initiative also transitioned a process from one that required multiple stakeholders to one that was essentially a self-service solution with few opportunities for defects.

Keep in mind that one of the key tenants of LSS is the reduction of all non value-added waste. It is difficult to eliminate waste if waste is not properly defined by the customer. If the customer is not involved as part of the improvement process, the ultimate outcome may cause future defects or concerns. If only a sample of the customer population can be involved as part of an initiative (due to size, scope, or resource constraints), an appropriate communication channel must be opened with all customers to ensure they are provided an opportunity to provide both positive and negative feedback. The key is to begin involving customers as part of the LSS program through process mapping, SIPOC, value analysis, and other improvement events. If customers are not involved on the front end, they are more likely to be unsupportive of any future improvement, implementation, or sustainment plans. See Table 2.7 for a list of approaches, tools, and methods used for increasing customer focus.

Table 2.7 Approaches, tools, and methods for increasing customer focus.

• Executive steering committee • Integrated stakeholder and communication plan • Strategic planning • Visible balanced scorecard • SWOT analysis	• Project selection • Project kickoff meeting • Project charter • SIPOC • Vo"X" • CTQs	• EWPM • Kaizen • Future state mapping • Benchmarking • Sustainment plan

HIGH EMPLOYEE TURNOVER

Going away parties seem to be one constant in the public sector. Whether due to promotions, internal position changes, military assignments and deployments, or changes relative to elected/appointed officials, there is always turnover. It can be difficult to maintain a robust LSS program when champions, trained personnel, and team members are constantly changing. New personnel must be trained, new champions identified, and projects rescheduled due to turnover (there's nothing like trying to replace the person who was responsible for the measurement plan).

One of the most powerful tools to limit the impact of turnover is an executive steering committee. (See Table 2.8 for other tools.) This provides ownership to positions rather than specific people, limiting the impact someone's departure has on the organization. One advantage (yes, I said advantage) to employee turnover is that it provides an opportunity for LSS principles to permeate throughout the organization. Someone who was excited about the tools and methods in a previous role will work to deploy and implement those same approaches in a new role, thus creating a ripple effect for the LSS program. A high turnover rate can be a concerning aspect during the deployment of a LSS program in the public sector, but a sound change management plan, stakeholder communication, and the sustainment of an executive steering committee can change the perceived negative impacts of turnover into a positive impact. The key to overcoming this opportunity is vigilance.

Table 2.8 Approaches, tools, and methods for overcoming employee turnover.

• Executive steering committee • Integrated stakeholder and communication plan • Risk management plan • Strategic planning • Visible balanced scorecard	• Funding reallocation Plan • Project charter • EWPM • SOPs • 5S	• Future state mapping • Control charts • Implementation plan • Sustainment plan

> **Public Sector Project Highlight**
>
> **Overcoming Consistent Turnover**
>
> One public sector organization decided to take a more direct approach to overcoming consistent turnover. They implemented a robust standard operating procedure for each key process, including the assignment of ownership for each process, process maps, detailed process descriptions, and the relationships to policy, laws and other regulations. This organization also took knowledge management to another level by deploying their operating guides across the globe via an intranet and integrating a continuous process improvement feedback loop. This allowed anyone utilizing the process to recommend changes or best practices to the process owner. The process owner could then decide whether changes add value to the process and, if so, deploy an updated operating procedure across the globe in real time.

COMPLEXITY OF THE PUBLIC SECTOR

So how is the public sector more complex than the private sector? First, there is no common focus such as keeping the business alive or increasing revenue. In the public sector, goals are specific to a particular department or agency. This can sometimes cause an overlap in mission and vision, as well as an overlap with customers. For example, multiple government contracting agencies can cause competing interests. Beyond the lack of common enterprise goals, there is separation of duties required in some areas of the government (such as between government agencies and Congress; there is no single "decider" on most LSS initiatives). There is good reason for some of these situations, such as the need to limit power entrusted to a single entity, but it can cause significant time delays when trying to accomplish a LSS initiative.

Another problem arises when there are competing interests between government agencies and the current administration. This can sometimes lead to an inability to obtain funding or can cause leadership changes in the form of presidential appointees.

The public sector is more complex than the private sector in many areas. One common problem caused by this is an inability to act strategically when it comes to LSS or other process improvement programs; in many cases funding is allocated on an annual basis and the focus of an administration can shift from year to year. Time delays are caused by the inability to make changes without approvals from external agencies. Although these can result in significant hurdles as part of any LSS initiative, they can be overcome if the tools listed as part of this section are applied correctly. (See Table 2.9.) One of the most critical aspects to overcoming a complex environment is getting all interested parties in a room. This allows the team to have an open dialogue about issues, concerns, or opportunities for improvement. I have never seen an agency that did not want to improve. They just required the opportunity to improve.

Table 2.9 Approaches, tools, and methods for overcoming complexity.

• Executive steering committee • Integrated stakeholder and communication plan • Risk management plan • Strategic planning • Visible balanced scorecard • Lean-Six Sigma audit • Affinity diagram • Project charter	• SIPOC • CTQs • EWPM • SOPs • Spaghetti diagrams • Kaizen • 5S • Future state mapping • Measurement plan	• Fishbone diagrams • Check sheets • Control charts • Histograms • Pareto charts • Scatter diagrams • Implementation plan • Sustainment plan • Run charts

Public Sector Project Highlight

Overcoming Complexity

The public sector is a large entity serving a multitude of customers, agencies, and organizations. For example, when completing a recent LSS initiative one team was tasked with streamlining a congressional reporting requirement. This process was completed by multiple agencies within a large public sector department. In order to increase the likelihood of gaining buy-in from Congress to make changes to the existing process, the LSS team sought input, insights, lessons learned, and other information from all agencies within the department. In order to overcome complexity, the LSS project team focused on stakeholder involvement and open communication. This collaboration not only resulted in successfully overcoming the complexity of the process and implementing a sustainable solution, but also enhanced the final recommendations providing for a 30% reduction in non value-added materials and a 33% reduction in process waste.

MIX OF EMPLOYEE TYPES

The private sector encompasses a mix of employee types: salaried and non-salaried, union and non-union. The public sector adds increased complexity to the mix of employee types with government and non-government, elected and non-elected, military and non-military, civilian and non-civilian, and the list goes on. This can cause issues from a variety of angles respective to LSS programs.

The first is that from an accountability perspective, the project champion may not have a direct ability to dictate the workload for specific resources. This can cause significant problems during the improvement, implementation, and sustainment phases of a project. Another problem may arise in implementing initiatives that affect a specific person's job role. A precise position description may exist to outline the requirements for a resource and specify what job functions that role is authorized to carry out. Changes proposed as the result of an improvement project may change a position description, requiring approval and input from labor unions, senior leadership, and other affected stakeholders (for example, personnel centers).

Public Sector Employee Types			
Military	Union	Contractor	Elected
Civilian	Non-union	Appointed	Term

Although there are a number of approaches for overcoming these difficulties, which will be discussed throughout this book, the one I have seen have the most impact is a robust executive steering committee. (See Table 2.10 for more resources.) In this instance, robust is defined as having various levels of leadership (that is, middle managers and executive leaders), as well as representatives from all possible employee segments (see table above) whether they be military, civilian, elected, appointed or contractor personnel. Everyone must have a voice and be treated as an integral stakeholder; otherwise, they will work to stop any unwanted improvements regardless of the impact.

Table 2.10 Approaches, tools, and methods for overcoming a mix of employee types.

• Executive steering committee • Integrated stakeholder and communication plan • Risk management plan • Strategic planning • Visible balanced scorecard • Funding reallocation plan • SWOT analysis • Project selection • Brainstorming	• Lean-Six Sigma audit • Affinity diagram • Multi-voting • PICK diagram • Project kickoff • Meeting • Project charter • SIPOC • Vo"X" • CTQs	• EWPM • Spaghetti diagrams • Kaizen • 5S • Future state • Mapping • Measurement plan • Fishbone diagram • Implementation plan • Sustainment plan

---

Public Sector Project Highlight

Bringing People Together

There is no better way to overcome a mix of employee types than getting everyone in a room together. That is why kaizen events are so powerful for creating a sense of teaming. These events begin with stating a common goal and then providing leadership support and the resources necessary to implement improvements quickly. As part of a recent LSS initiative, one team realized that the final improvement recommendation hinged upon gaining support across the entire enterprise, which consisted of multiple employee types. In order to gain this support, the LSS team decided to facilitate an accelerated kaizen event focused on reviewing the project charter, the value stream analysis, and the improvements identified as part of the value stream effort. The team then ensured everyone was provided an adequate opportunity to voice concerns and provide input into the final implementation and sustainment strategy. Although the improvements seemed straightforward, the LSS project team realized sustainment would be nearly impossible without support across all the employee segments.

Section 3
Laying the Groundwork for Success

"The achievements of an organization are the results of the combined effort of each individual."

– VINCE LOMBARDI

In the construction business, everyone knows that a good house cannot be built without a solid foundation; using LSS in an organization follows the same guiding principle. In this section we will review the fundamentals required to ensure success. This includes level-setting with leadership, managing employee and personal expectations relative to LSS results, identifying key project team members and stakeholders (as well as a communication approach for each of them), managing and mitigating potential risks, developing the organization's overall strategic plan, mission, vision, goals, and metrics, creating a funding re-allocation plan for any savings gained, and identifying possible high-value initiatives. I will also provide you with the top reasons LSS programs fail. If all of these items are appropriately discussed and planned for, there will be a solid foundation for success.

EXPECTATION SETTING

There is nothing more disheartening than having a LSS project killed by senior leadership because it seems to cost too much or is moving too slowly or because results are not being realized. As you gain initial buy-in to run a LSS project, which should be done during the project charter process as described later in this book, the LSS team lead must ensure the timeline, goals, and business case are formally agreed upon and signed off on by the project champion. If this step is not completed, your project may be stopped mid-stream. This will ultimately affect your ability to gain stakeholder buy-in and support on future LSS initiatives.

KEY PROJECT STAKEHOLDERS AND TEAM MEMBERS

Change management is critical to the success of any LSS program. One of the first steps in a good change management plan is understanding who the key players are. Below is a list of key stakeholders involved in a LSS project and how they can impact the overall success of your program.

Executive Leadership and Deployment Champions

These are the "make or break" players in any LSS project. Leadership buy-in is critical to the success of any improvement program; special attention should be given to ensuring this buy-in exists. Executive and project champions should attend executive leadership training and learn how they can impact and foster an environment of collaboration, eliminate organizational barriers, and gain organizational support. They will be the ones who provide the resources; the LSS project team is responsible for using these resources in an efficient and effective manner, ensuring a positive impact on the organization.

Process Owners

Process owners are responsible for the output of the process. In a manufacturing environment, this may be the line manager or supervisor; in a services environment, this may be the individual who is responsible for the department or a specific reporting requirement. It is essential to obtain support from the process owner early on in the project, targeting buy-in when developing a project charter so that the process owner can enable overall success and support the change management aspects to implement any identified improvements. These individuals should at a minimum attend an awareness level LSS training class.

Public Sector Project Highlight

Importance of Process Owners

Based on the stovepiped environment of the public sector, it is essential to ensure involvement and buy-in from the process owner in advance of starting the LSS project. If process owners are not involved in defining the problem, analyzing the opportunities, identifying improvements, and developing the long-term sustainment plan, it is unlikely they will be supportive of the change. Leveraging previous LSS results, a robust change management plan, and the support from the executive steering committee, always engage the process owner and gain support before moving forward with a project charter.

LSS Master Black Belts

Master Black Belts or MBBs are expected to be experts in all aspects of the LSS methodology. They must have expert knowledge of advanced statistical tools, the ability to deploy the methodology across a large-scale enterprise, the capability to set both short-term and long-term improvement goals and objectives with relative performance measures, and an ability to lead enterprise-wide, highly complex projects. One of the most important roles of an MBB is the proven ability to train, coach, and mentor individuals ranging from Black Belts to the executive Leadership of the organization. An MBB is a full-time, dedicated resource focused on the deployment and sustainment of the organization's LSS program.

LSS Black Belts

Black Belts or BBs are experts in all of the LSS tools and lead several complex initiatives each year. A typical BB will complete four to six initiatives annually, netting a $1M+ impact to the organization's bottom line either via cost savings or cost avoidance. They may also mentor and train Green Belts (GBs) across the organization. Similar to the MBB, this is also a full-time role.

LSS Green Belts

Green Belts or GBs are individuals who lead less complex initiatives and work under the supervision of either a BB or MBB to complete a project. These individuals support LSS programs on a part-time basis in addition to normal responsibilities.

LSS Yellow Belts

Yellow Belts are individuals who are not responsible or expected to lead initiatives, but who have completed an overview LSS DMAIC training including some basic LSS tools (typically two days in length). This role is instrumental from a change management perspective in helping the organization understand the approach and why these projects are being implemented.

LSS White Belts

White Belts are individuals who have completed awareness training typically lasting two to four hours in length and covering the LSS DMAIC approach at a high-level. The purpose of this training is to begin to permeate an organization with the value of the LSS approach. A good target for LSS White Belt certification is 100%. One method to achieve this goal is via lunch-and-learn sessions.

Although this training is sometimes skipped over, it is critical to help institutionalize the LSS methods and eliminate fear of change. This will help foster an environment of openness where team members understand that they are not a target but rather a support function in helping to make the organization better as a whole. LSS aims to empower individuals and provide an opportunity for all personnel to voice their concerns. This can be more easily accomplished if people understand the LSS methods.

Functional Subject Matter Experts

A GB, BB, or MBB will not be able to complete a LSS initiative without an intimate understanding of the current product, process, or service. This is where involvement from subject matter experts (SMEs) becomes critical. Their input is instrumental in the identification of valid improvements and they are essential from a change management perspective. If you gain their insights and listen to their frustrations and concerns on the front end, it will make implementation and sustainment much easier. These individuals should at a minimum complete awareness level training.

Accounting and Finance

An improvement is not complete unless it is validated. Accounting and Finance should be involved in part of this validation. If a project deals with saving time, paper, or other basic resources, Accounting and Finance will be able to provide the cost of these items. If it deals with savings attributed to warehouse space, they will be able to provide the cost savings (the list of information available from Accounting and Finance continues). Beyond validating improvements, Accounting and Finance typically maintains a direct line to the most senior levels of leadership. If Accounting and Finance personnel begin to see significant LSS savings, they will help carry the "buzz" across the enterprise.

EXECUTIVE STEERING COMMITTEE

Impact in Overcoming Public Sector Challenges									
Breaking down stovepipes	Creating urgency	Leadership support	Metrics	Common goals	Increasing customer focus	Reducing impact of turnover	Overcoming complexity	Fostering collaboration
X	X	X	X	X	X	X	X	X

As part of every LSS program, there should be an executive steering committee focused on ensuring there are metrics tied to training, the cost vs. benefit of the initiatives, the number of projects completed, and other performance measures key to the sustainment of a robust LSS program. This committee should be a representative mix of all stakeholders including members from executive leadership, key process owners,

Lean-Six Sigma Belts, SMEs, and Accounting/Finance. This committee should have a set meeting schedule (for example, monthly) and should communicate milestones obtained or change in vectors for the organization's overall LSS program.

The executive steering committee is charged with eliminating roadblocks for the LSS program. This includes:

- Fostering a collaborative, non-attribute environment where employees feel open to submitting opportunities for improvement,
- Identifying champions for each LSS effort and providing them with the required resources to complete each project, and
- Ensuring that a robust governance structure is in place including goals, metrics, improvements, and monthly status updates presented to leadership.

The simplest approach to ensuring support from the executive steering committee throughout the LSS process is to clearly link their leadership indicators (performance reviews) to the results achieved as part of the LSS program. This will provide for a robust governance structure in the near term and excitement and buy-in for the long-term LSS strategic plan.

Public Sector Project Highlight

Training is Required for ALL Stakeholders

Lean-Six Sigma is systematic in its approach to solving problems and sustaining improvements. LSS has evolved into a systematic approach for training belts, champions, and all other stakeholders involved with the LSS program. In the public sector, special attention should be paid to ensure that everyone obtains some level of training relative to the LSS methods and tools and should participate in a simulation demonstrating how quickly results can be obtained. The key is to engage your executive steering committee early to develop a training approach. This must include sponsor and executive training, belt training and most importantly the development of a mandatory awareness-level training class. Training in this case is really part of the overall change management strategy. People who are not aware of the value provided by LSS will be less likely to buy in to future initiatives.

INTEGRATED STAKEHOLDER AND COMMUNICATION PLAN

Impact in Overcoming Public Sector Challenges								
Breaking down stovepipes	Creating urgency	Leadership support	Metrics	Common goals	Increasing customer focus	Reducing impact of turnover	Overcoming complexity	Fostering collaboration
X		X		X	X	X	X	X

One of the biggest obstacles to utilizing LSS within a public sector organization is the fear of change. Change management is critical during deployment and implementation and throughout each and every LSS initiative. A fundamental requirement for a sound change management plan is a clear understanding of the stakeholders involved and a method to keep them informed.

From a stakeholder identification perspective, you want to take the input gathered during the development of your project charter and drill down to the lower level functional organizations that will require information. An excellent tool to begin this process is an integrated communication plan and stakeholder analysis. From a stakeholder analysis perspective, there are five key steps:

1. Brainstorm all possible stakeholders involved in your LSS initiative. Include organizations that are directly impacted and also organizations that may be indirectly impacted and therefore in need of information.
2. Identify the impact of the process on each stakeholder using the scale indicated in Table 3.1.

Table 3.1 Impact of the process/project on stakeholder.

High = 9	Direct and significant impact to operations of an office or individual.
Medium = 3	Potential or indirect impact on the operations of an office or individual.
Low = 1	Minimal or no impact on the operations of an office or individual.

3. Next identify their power to transform the process using the scale indicated in Table 3.2.

Table 3.2 Power to transform the process/project.

High = 9	Decision-maker regarding how the process works.
Medium = 3	Provides input to the decision-maker.
Low = 1	Minimal input on how the process works.

4. As part of each stakeholder analysis, baseline everyone's current commitment to the initiative as indicated in Table 3.3.

Table 3.3 Commitment to process/project change.

Uncommitted = 9	Against the change; must be completely convinced in order to gain buy-in.
Committed with reservations = 3	Slightly committed or undecided; must be persuaded to be fully committed.
Completely committed = 1	Already committed without any need for extensive persuasion.

5. The last step involved in the stakeholder analysis part of your integrated plan is to identify stakeholder levels of expertise using the scale indicated in Table 3.4.

Table 3.4 Level of expertise on the process/project.

High = 9	Subject matter expert.
Medium = 3	Foundational understanding of how the process works.
Low = 1	Limited or no knowledge of how the process works.

Based on your understanding of each stakeholder and the ratings determined above, identify an appropriate communication strategy. Multiply your ratings for each stakeholder and rank them in terms of the highest numerical value. The initial focus and effort should be placed on the stakeholders with the highest values, but every stakeholder identified should have some sort of communication tied to them. In developing a robust communication plan, consider the elements found in Table 3.5.

Table 3.5 Communication strategy.

Key message	What is the message that must be delivered to the stakeholder (for example: obtain a decision, provide input, informational purposes, and so on)?
Vehicle	How will you deliver your message (for example: email, phone, formal meeting, article, conference, and so on)?
Frequency	How often will you deliver your message to a particular stakeholder (for example: real-time, daily, monthly, quarterly, annually)?

Keep in mind that the completion of this simple document can significantly impact the success of the overall LSS project. If stakeholders are not identified on the front end, involved, and kept abreast of changes that may impact their organization, it may be difficult to obtain buy-in when it comes time to implement. A robust, integrated stakeholder management and communication plan (see Figure 3.1) is a simple way to visually depict how everyone impacts the initiative and how they will be kept informed.

Laying the Groundwork for Success 39

Project ABC – Communication Plan/Stakeholder Analysis											
Goal	There is a great need to communicate to stakeholders the importance of Project ABC from an end-to-end perspective. We must determine a method to saturate the organization from both a horizontal and vertical integration perspective (such as top-down and across).										
	Relationship to project						Communication/involvement strategy				
Stakeholder	Impact of process on stakeholder	Power to transform process	Commitment	Level of expertise	Overall rating		Meet with regularly	Value stream mapping event attendees	Monthly organizational newsletter	Conferences	Send copy of meeting minutes	Speak with informally as needed
					Notes	Frequency	Weekly	Ad-hoc	Monthly	Quarterly	Ad-hoc	Ad-hoc	Notes
Manufacturing	9	3	3	9	729		X	X	X	X	X	X	
LSS project team	9	3	9	1	243		X	X	X	X	X	X	
Procurement	1	1	1	3	3			X	X	X	X	X	
Finance and accounting	1	3	1	1	3					X		X	X
Human resources	1	1	1	1	1					X		X	X

Figure 3.1 Sample communication plan/stakeholder analysis.

RISK MANAGEMENT PLAN

Impact in Overcoming Public Sector Challenges									
Breaking down stovepipes	Creating urgency	Leadership support	Metrics	Common goals	Increasing customer focus	Reducing impact of turnover	Overcoming complexity	Fostering collaboration
	X					X	X	X

As part of every LSS effort there will be some inherent risk relative to project timeline, scope, and resources or specific to the improvement opportunity (such as safety, readiness, and legal/regulatory requirements). The intent of facilitating a risk-related discussion as part of the LSS program and initiative involves the determination of several important issues:

- The risks that could occur.
- The potential impact of those risks.
- The probability of each risk occurring.
- The stakeholders who would be impacted.
- The severity of risk occurring.

In facilitating this type of review, the first step is to use brainstorming to identify potential risks. Then, using expertise from the organization subject matter experts and project team, determine what the possible impact would be if each risk actually occurred. Would it affect the customer (for example, a defect), the supplier (for example, a late payment), or someone internal (for example, a safety issue)? For ranking purposes later, it's a good idea to obtain a generic ranking for the impact of each risk at this point (high, medium, or low).

When you have a list of risks, the possible impacts to specific stakeholders, and a generic ranking for each, identify the probability of each risk occurring. I recommend utilizing a percentage range for this step (for example, 0-20%, 21-40%, 41-60%, 61-80%, 81-100%). As a final step in each risk management session, assign an appropriate owner to each risk, someone who will be in a position to either see the risk about to occur or see the risk actually occur. The inputs gained throughout the risk management session can now be plotted into a matrix similar to the one illustrated in Table 3.6. This example only displays one risk, but as part of a risk management session, the outcome could be any number of potential risks.

Table 3.6 Example risk management plan.

Priority	Risk description	Potential impact	Probability	Severity	Stakeholders	Owner
1	Unable to submit timecards before the end of the work week.	Inability to process payroll due to lack of employee timekeeping data.	0-20%	H	All internal employees	John

The trickier step to risk management is deciding whether a reactive or proactive approach is more appropriate:

- Is the proposed risk a one-time occurrence or can it happen multiple times?
- Is the severity of the risk so high (as illustrated in Table 3.6), that the risk of occurrence, although small, is still too high?

This should be an informed decision based on all the information collected regarding probability and impact. An appropriate response should be prepared for all risks and a member of the team should be assigned to continuously monitor each risk as appropriate. (See Table 3.7.)

Table 3.7 Example risk mitigation plan.

Priority	Risk description	Potential impact	Mitigation
1	Unable to submit timecards before the end of the work week.	Inability to process payroll due to lack of employee timekeeping data.	If employees are unable to report their timekeeping information electronically, they will submit manual timecards to their supervisors. Each supervisor will be responsible for reporting the data to payroll directly.

Upon completion of any LSS initiative, it is beneficial to publicly communicate all risks identified, any mitigation strategies used, and other lessons learned to help support future LSS efforts. The risk matrix should also be provided to the executive steering committee for review. The organization should not fear proactively identifying risks. Every project involves some inherent risk. The purpose here is to identify, rank, and eliminate or mitigate. It is through continuous data sharing that an enterprise-wide LSS program will be successful.

STRATEGIC PLANNING

Impact in Overcoming Public Sector Challenges								
Breaking down stovepipes	Creating urgency	Leadership support	Metrics	Common goals	Increasing customer focus	Reducing impact of turnover	Overcoming complexity	Fostering collaboration
X	X	X	X	X	X	X	X	X

Before we discuss what is involved in the strategic planning process, we must first understand the value provided by this process. Successfully completing a strategic planning session with senior leadership is the foundation for continued buy-in and support across the entire enterprise. It should define what your organization does (mission), who you do it for (stakeholders), and how you plan to excel at your core value proposition in the future (vision). A strategic plan can also integrate other items such as core values. Figure 3.2 displays the relationship between a LSS strategy and execution of that strategy, with a continued focus on your true mission and vision.

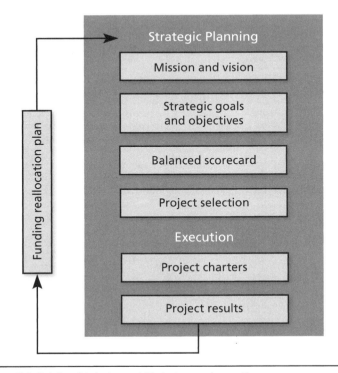

Figure 3.2 Relationship between LSS strategy and execution.

The first step in the strategic planning process is to identify your mission and vision statements. If you are working within a subcomponent of the overarching organization, these should be clearly linked to the organization's overall mission and vision (see Table 3.8).

Table 3.8 Example mission and vision statements.

Mission	Provide customer service in a timely and accurate manner to every citizen. This will be achieved by hiring/retaining the best customer service personnel, as well as leveraging best-in-class technology to enhance the customer experience.
Vision	Premier service provider to all citizens.

A **mission statement** defines the purpose of your organization. It should be used to guide your organization's work to ensure you provide your stated value to all stakeholders. This should clearly link to your vision statement, which defines your long-term view of the value your organization will provide in the future.

The **vision statement** should clearly articulate where your leadership would like to take the organization in the long term or what value they would like to provide to their stakeholders.

In order to obtain an organization's mission and vision statement, the following approach should be used:

- The first step in facilitating a successful strategic planning session is to have the appropriate attendees in the room representing the entire organization. If you will be addressing procurement, public relations, and budgetary aspects, you should ensure senior level champions, as well as a representative mix of middle managers and front-line workers, are in the session. Having the appropriate attendee mix will ensure that your strategic plan maintains a focus on the holistic charter and prevent unnecessary overlap between goals, objectives, and metrics.
- When you have the right people in the room, you must ensure that everyone understands the ground rules. The most critical rule is that everyone has the same "voice." Regardless of rank or seniority in the organization, everyone should feel comfortable providing input in a non-attribute environment. What this means is that all discussions as part of the strategic planning session, even if they seem negative towards leadership or other organizations, will not transfer out into the operations of the organization (for example, in performance reviews). Other ground rules that support the facilitation of a successful session include leaving all electronic devices at the door, eliminating the word "can't" from

the available vocabulary, and trying to create an environment of collaboration, openness, and trust.

- Facilitate a brainstorming session and list what the organization provides currently and what it wants to provide in the future. Ignore how this will be provided at this point; concentrate on what the organization provides (this will be achieved as part of the SMART Goals and Objectives section).
- Use an affinity diagram to categorize this output. This can be used to draft mission and vision statements.
- Multi-vote if necessary to determine the final mission and vision statement. Then publicize it across the organization internally and distribute it to all possible external sources as well. This will help to solidify what an organization provides to other organizations in terms of value.

Remember that a successful mission and vision statement is a direct correlation to your overarching mission and vision. The highest level of an organization must have a mission and vision statement, as well as the lowest sub-organization. This helps to ensure continuity when pursuing strategic goals and objectives. It also supports an environment of leadership buy-in and organizational support. If the definition for your sub-organization does not match the overall stated mission or vision, then you may not be adding value.

SMART GOALS AND OBJECTIVES

When working to define your goals and objectives, you must first understand the difference between the two. A goal is your broad desired state; an objective is a tangible action to reach your stated goals. See Table 3.9.

Table 3.9 Example goal and objectives.

GOAL
Decrease overall personnel costs (measurable) by 10% (specific/attainable) within the operations group (relevant) by December 31st, 2011 (time bound).
OBJECTIVE 1
Decrease adminstrative personnel costs (measurable) by 25% (specific/attainable) within the operations group (relevant) by December 31st, 2011 (time bound).
OBJECTIVE 2
Decrease maintenance personnel costs (measurable) by 5% (specific/attainable) within the operations group (relevant) by December 31st, 2011 (time bound).

Based on your mission and vision statement, work with senior leadership to define SMART goals and objectives. SMART is the acronym meaning specific, measureable, attainable, relevant, and time bound. See Table 3.10.

Table 3.10 SMART goals.

	SMART
Specific	The goal, objective, and/or metric must be a quantifiable target, (for example, a 20% reduction).
Measurable	A sound goal, objective, and/or metric must be defined in terms of success (for example, cost savings, cost avoidance, cycle time, or other efficiency/effectiveness measures).
Attainable	Stretch goals are great, but in order to gain support and buy-in the goal, objective, and/or metric must be something that can be obtained. For example, obtaining $100 million in cost savings in an organization that has a total budget of $120 million may not be appropriate. Otherwise the project team may have negative feelings.
Relevant	The defined goals, objectives, and/or metrics must be somehow tied to the organization's overall mission/vision. For example, if the mission of an organization is to provide the highest levels of customer service, reducing customer facing time may not be relevant.
Time bound	Goals, objectives, and/or metrics do not add value if they are not timely. In this case, it means putting a target on when goals and objectives will be achieved (for example, by the end of the fourth quarter of the current fiscal year or within the next 12 months).

It is critical that your goals, objectives, and other project metrics meet these five criteria. Once you have worked with your organization to identify your goals and objectives, you can create detailed metrics to support the completion of stated objectives.

VISIBLE BALANCED SCORECARDS AND METRICS

Impact in Overcoming Public Sector Challenges								
Breaking down stovepipes	Creating urgency	Leadership support	Metrics	Common goals	Increasing customer focus	Reducing impact of turnover	Overcoming complexity	Fostering collaboration
X	X	X	X	X	X	X	X	X

A balanced scorecard is a metric dashboard that not only clearly shows all relevant metrics to achieve the organization's strategic plan, but also ensures a balance between long-term goals, financials, customers, and employees. Visible and balanced scorecards provide the foundation for all strategic, tactical, and improvement discussions. How can one scorecard support this entire decision-making process? That is easy, since it has linkages to the organizational mission, vision, and strategic goals and objectives and it is balanced and constantly visible to all personnel.

The key to a robust scorecard is that it is visible and balanced. Visible means that people in the organization can see how they are doing relative to their individual metrics. This can be achieved in a simple fashion by using a corkboard with key metrics updated on a daily, weekly, or monthly basis. A more advanced version would make use of a web-based or digital dashboard that displays execution on a real-time basis. This allows an organization to react to real-time data, but the more simplified approach will work for the near term as you turn the corner on your LSS journey.

The second part of the equation is making sure that the information displayed is balanced. It should include information about financials, customer satisfaction, employee retention and training, and the achievement of your mission and vision. (The initial criteria developed by Kaplan & Norton for a balanced scorecard focused on four areas: financial, customer, internal business processes, and learning/growth). A balanced scorecard should be constructed using SMART criteria and the final outcome should have clear linkages to the mission, vision, and core competencies for your organization.

So now that you understand what a visible and balanced scorecard encompasses, it's time to build one. Since the foundation for all organizational metrics should be the mission and vision statement, the easiest approach is to include it as part of your strategic planning session.

Follow these steps to construct your scorecard:

1. Using the mission, vision, core competencies, goals, and objectives identified as part of your strategic planning session, identify metrics or success criteria for each area.
2. Now that you have identified metrics, use an affinity diagramming approach to clearly link all metrics to a financial, customer, employee, or organizational vitality impact area.
3. Next construct the initial balanced scorecard (see the theoretical example provided in Figure 3.3), ensuring that the mission, vision, goals, objectives, and metrics are included.

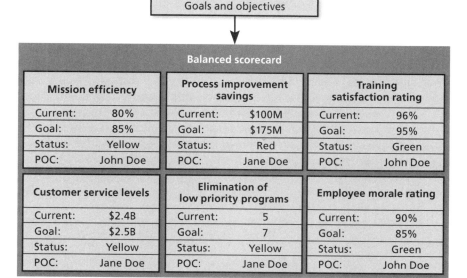

Figure 3.3 Theoretical balanced scorecard.

Beyond the clearly identifiable success metrics (such as reduced waste), there should also be clear deployment, personnel development, and other LSS metrics (for example, number of personnel trained in LSS and number of projects completed). A list of example metrics includes the following (this is a good list to start the metric brainstorming process with your executive steering committee and organizational stakeholders):

- Customer service
- Waste
- Number of personnel LSS trained
- Total cost savings
- Cycle time
- Inventory levels
- Machine up/down time
- Defects/defectives
- First-time pass yields

4. Finally, ensure buy-in from leaders within the organization and create/deploy a robust communication strategy to ensure everyone understands the new metrics, the impact on them, and where a current status can be seen. Assign each metric to a point-of-contact and organization. This helps establish clear success criteria for every person in the organization. Make sure that there is a sustainment plan for the scorecard to ensure the metrics remain focused on valuable aspects of the organization (the strategic plan). The worst thing that can happen to a scorecard is that it loses its relevance. If one of the metrics is trending too negatively (that is, to yellow or red), it must trigger a reaction (improvement plan, action plan).

A robust scorecard becomes the evaluation criteria for LSS project selection and prioritization, while also providing a common focus for the entire organization. Some of the most influential leaders I have seen in both the public sector and the private sector carry a print-out of their balanced scorecard with them at all times. If something is not part of the balanced scorecard, then it must not be of primary importance to the organization.

> **Public Sector Project Highlight**
>
> **Visible Balanced Scorecard**
>
> In facilitating a recent LSS initiative, we were required to translate 15,000 line items of data into key improvement areas. Although we were able to commit a small team of four people over a period of five months to analyze this data, we realized this would not be possible during the sustainment. We needed to develop a scorecard that provided real-time, visible information to the organization in order to react to any changes. This was deployed through an automated IT dashboard, which pulled real-time data from key systems without any human interaction. The result was key metrics to sustain an initiative that provided millions of dollars in annual cost avoidance.

CREATE A FUNDING RE-ALLOCATION PLAN

Impact in Overcoming Public Sector Challenges									
Breaking down stovepipes	Creating urgency	Leadership support	Metrics	Common goals	Increasing customer focus	Reducing impact of turnover	Overcoming complexity	Fostering collaboration
X	X	X	X	X		X		X

Special care should be taken to ensure that a funding re-allocation plan is in place prior to starting a public sector LSS initiative. Earlier in the book we discussed issues that can arise when an organization is not profit focused. Having a funding re-allocation plan will help overcome this shortfall by ensuring that savings gained will be immediately used for another mission-critical requirement. (See Table 3.11.)

Table 3.11 Funding re-allocation plan.

Mission critical requirements	Cost	LSS projects	Savings
New computers	$2.4M	Energy and utilities cost savings initiative	$30M
Updated cyber security software	$25M		

Since the goals and objectives have already been defined as part of the strategic planning process, these will be used as the baseline for projected efficiency or effectiveness gains. With those measures as the framework, a formal discussion should be facilitated with the organization's executive steering committee and any key leaders not involved in the committee to identify a prioritized list of unfunded but mission-critical requirements. The tools used for this discussion will be brainstorming and multi-voting, both of which are discussed later in this section. The items identified can then be tied to the appropriate measures. The final step in the process is to obtain a formal signature from all involved senior leaders to gain sign-off and assure that project successes will be immediately re-infused to support mission-critical efforts. Doing this gives every project a purpose in the public sector, similar to every project having an impact on the bottom line in the private sector.

In the theoretical example provided, the cost savings would be immediately re-infused to purchase new computers and update cyber security software. It is critical that leadership continues to be supportive and provide updates to the funding re-allocation plan. If the savings are not re-infused for mission-critical requirements, the overall success of the LSS program could be at risk.

> **Public Sector Project Highlight**
>
> **Funding Re-allocation Plan**
> The key to gaining funding support for initiatives in the public sector is understanding what savings can be gained from completing the project and also how mission-critical funding can be re-infused in an increasingly constrained environment. Congress goes through these iterations during every budgetary cycle as part of appropriating public funds. It is the responsibility of the steering committee, with input from the organization and the LSS team, to provide the same level of fiduciary control and oversight.

TAILORED APPROACH

As with most things, there is an ideal approach to implementing and using LSS and then there is reality. A few approaches outlined here can help you overcome potential roadblocks in the public sector while reaping the benefits LSS projects provide.

The first tip for tailoring the methodology is to keep it simple. This means dropping or tailoring the terminology for specific organizations. For example, in the public sector I have often heard kaizen events referred to as rapid improvement events. Although the terms are slightly different, the intent, approach, and result are similar—providing rapid changes to an existing product, process, or service.

Another aspect of keeping it simple is to use the simplest tool to achieve the desired outcome. If a scatter diagram will provide enough information about the correlation between two variables, you don't need regression analysis. A common misconception is that more advanced tools will produce a better outcome. In my experience, the inverse has been true. Project stakeholders may not buy in to the final result if they don't understand the analysis, and the project may fail altogether if leadership does not buy in to the methods.

Beyond keeping it simple, give strategic thought to how LSS resources will be integrated as part of the organization. For example, in the private sector LSS Black Belt resources are typically focused on completing projects as a full-time job. It's not necessary to follow the same approach in the public sector, but it's important that potential belt candidates are aware of long-term roles and responsibilities and career progression opportunities. If it is feasible, it's a good idea to include a process improvement or quality function as part of the organization, in order to ensure an appropriate focus is given to continuous process improvement. It has been my experience that Black Belt and Master Black Belt resources should be completely dedicated to process improvement, both from a career progression and job responsibility stand-point.

From a training perspective, an organization must ensure that training is customized for the specific environment. If the environment is transactional, the training should focus on using LSS in the transaction-focused world. Executive, Sponsor, and Champion training should also include simulations that are clearly relevant to their organizations. Awareness training, which typically includes an overview of LSS terms (for example, DMAIC, lean, Six Sigma, kaizen, and so on) and a simulation must be provided to everyone in the organization. This will help to foster a community of common understanding as well as organizational buy-in.

WHERE DO I GET MY PROJECTS?

If your organization has not fully deployed LSS across the enterprise, you may not have a stack of projects waiting to be completed. Because the primary purpose of this book is to make an immediate impact in a public sector organization, I will provide a quick way to identify a list of potential high-value initiatives within your organization.

SWOT Analysis

Impact in Overcoming Public Sector Challenges									
Breaking down stovepipes	Creating urgency	Leadership support	Metrics	Common goals	Increasing customer focus	Reducing impact of turnover	Overcoming complexity	Fostering collaboration
X	X	X		X	X			X

Prior to going into your project selection meeting you should have an understanding of your organization's overall strengths, weaknesses, opportunities, and threats. (See Table 3.12.) A SWOT analysis allows you to visually assess where your organization has opportunities to improve at a high level. The easiest way to gather this information is to interview your key stakeholders and gain their perspective on the organization.

As you are developing your SWOT analysis, keep in mind that it is not necessary to improve all weaknesses. From a strategic perspective, some of your weaknesses may be areas your organization has deemed as non-core business functions or processes.

Table 3.12 SWOT analysis.

Threats	Strengths
EXTERNAL threats to your organization such as government mandates or competitors gaining a significant advantage.	**INTERNAL** areas in which your organization excels; areas in which you are known as a best practice.
Opportunities	**Weaknesses**
EXTERNAL areas in which your organization has an opportunity to expand, increase customer value, or increase revenue; any other opportunities that could positively impact your organization.	**INTERNAL** areas in which your organization is not excelling; areas that could use some improvement.

Project Selection Meeting

Impact in Overcoming Public Sector Challenges									
Breaking down stovepipes	Creating urgency	Leadership support	Metrics	Common goals	Increasing customer focus	Reducing impact of turnover	Overcoming complexity	Fostering collaboration
X	X	X	X	X	X			X

A key component of any successful LSS deployment is proper project selection. Think of it as purchasing a new vehicle; if you test drive a car and it does not run well, you will probably never buy that particular vehicle. The same is true with LSS and your leadership team. If they see mediocre or poor results in the first wave of projects, they will most likely never buy in completely to the methodology. It is essential to ensure that your first wave of projects makes an impact and that the results are sustainable.

Follow these steps to facilitate a successful project selection meeting:

1. Determine criteria for ranking your projects. This could be something as easy as decreased cost, increased capacity, or higher customer satisfaction levels. These should be quantitative whenever possible in order to eliminate guess work from a potential impact perspective. The criteria must be supported by all levels of leadership and linked to your strategic plan.

2. Identify stakeholders who must be involved as part of the meeting. The attendees should represent a cross section of all areas that may be impacted (management and lower-level personnel).

3. As part of the meeting, it is your role as the facilitator to ensure communication channels are open. If a certain person, even someone in management, is stifling the conversation, he or she must be removed from the meeting.

4. Develop an agenda for the first meeting along the lines of the one outlined here:

    Project Selection Meeting Agenda:

    Opening Remarks – This is where your champion or sponsor should open the communication channels and make it clear that this is a non-attribute environment.

    Meeting Purpose – Describe the overall purpose of the meeting and explain why your organization must transform; assure attendees that the meeting will focus on opportunities for improvement, not on "personnel type" items.

Criteria Review – Review the criterion that has been approved by leadership. Inform attendees that projects should be identified based upon this criteria.

Brainstorming – This should be an anything/everything brainstorming session to identify areas for improvement. If you are dealing with a large group (more than eight people), use break-out groups with facilitators. It is your responsibility to ensure this session is focused on generating high-level ideas. The remainder of your session will drive these ideas into more detailed project charters.

Impact and Goal Statements – This is similar to earlier enterprise goal and objectives discussions. Ask the group to provide details on the potential impact for each initiative. Keep the attendees focused on using the criteria provided by leadership.

Scope – Focus on identifying the start and the end of the opportunity you are attempting to fix.

Stakeholders – The intent during this section should be on identifying the process owners, suppliers, and customers for the improvement opportunity.

Multi-voting – Use the multi-voting technique, as described later in this chapter, to identify your top improvement areas.

Next Steps – Provide high-level next steps and communication vehicles to keep everyone informed of which projects are selected, how they are progressing, and where the meeting attendees may be asked to provide further clarification. This is the first part of your change management plan, keeping people involved.

5. Out-briefing leadership and gaining buy-in for your improvement initiative is critical. When displaying initiatives to senior leadership, consider using a PICK diagram. This puts the information into an easy-to-understand context. Ensure that all tools you plan to use throughout your LSS journeys are included as part of your executive/champion training.

At one time shampoo bottles came with instructions: Lather, rinse, repeat. The final part of any successful project selection effort will benefit from those same guidelines. Once you have completed your initial high-value initiatives, reconvene the project selection group and gather more ideas for improvement. This should be an ongoing cycle if you are committed to fostering an environment of continuous process improvement.

> **Public Sector Project Highlight**
>
> **Projects Can Be Identified Quickly**
>
> Using the key strategic improvement objectives as defined with senior leadership, we facilitated a four-hour project selection meeting. The result of this session was the identification of 28 validated improvement areas (Six Sigma, kaizen, quick-wins), with input gathered from a 35-person organization. These 28 initiatives were further defined in a limited timeframe of four hours to include: respective prioritization, problem and opportunity statements, scope, high-level stakeholder identification, and goal statements. The key to any project selection workshop is accurately mapping projects to key strategic improvement objectives, which should also be appropriately linked to the organization's strategic plan.

Brainstorming

Impact in Overcoming Public Sector Challenges								
Breaking down stovepipes	Creating urgency	Leadership support	Metrics	Common goals	Increasing customer focus	Reducing impact of turnover	Overcoming complexity	Fostering collaboration
X				X				X

Brainstorming involves the gathering of *all* possible ideas without judging. In this case the focus will be on identifying opportunities for improvement. The key to facilitating a successful brainstorming session at this phase is to foster a non-attribute environment. What this means is that any opportunities identified for improvement will not be looked upon as wasted effort or an attack against leadership. That is, leadership will not take any personnel actions (reprimand employees) based on ideas identified.

Another important component of a solid brainstorming session is an open environment where everyone feels "heard." If this does not occur spontaneously, a facilitator can go around the room and ask each person to suggest an idea. This will ensure that people who are reluctant to speak have opportunity to contribute. Brainstorming sessions provide value to the LSS goals beyond identifying improvements; they also help "break the ice" in using LSS in a public sector organization.

Lean-Six Sigma Audit Worksheets

Impact in Overcoming Public Sector Challenges								
Breaking down stovepipes	Creating urgency	Leadership support	Metrics	Common goals	Increasing customer focus	Reducing impact of turnover	Overcoming complexity	Fostering collaboration
X	X	X	X	X			X	X

In order to continuously improve your overall LSS program, you need a method to analyze program maturity. The easiest way to accomplish this objective is to use LSS audit worksheets. The output from these worksheets will provide insights into areas for improvement across the LSS program and an overall maturity rating (high, medium, or low). The rating, as well as the areas for improvement, should be briefed as part of each executive steering committee meeting.

Questions can be tailored slightly depending on the focus, goals, and structure of a given organization. See Table 3.13 for an example LSS audit worksheet.

Table 3.13 Example LSS audit worksheet.

Training	Yes	No
Is there an adequate pipeline of candidates for the next six months of planned training?		X
Is a standardized training curriculum developed and deployed?	X	
Are simulations tailored for our specific environment?		X
Have all employees completed awareness training?		X
Has a robust certification process been implemented?		X
Project Results	**Yes**	**No**
Have all projects been completed within the anticipated timeframe?	X	
Have the results been sustained and validated?	X	
Have savings been re-infused per the funding re-allocation plan?	X	
Are less than 5% of the total number of projects in a "troubled" status?	X	
Are all initiatives tied to the strategic plan?	X	

(Continued)

Table 3.13 Example LSS audit worksheet. *(Continued)*

Sustainment	Yes	No
Has a visibile, balanced scorecard been developed and deployed?	X	
Have an executive steering committee and a frequent meeting schedule been established?	X	
Has the strategic plan been reviewed in the last quarter to ensure it is still relevant to the organizational mission and vision?	X	
Is there an appropriate pipeline of future initiatives (based on training plans and number of certified belts)?	X	
Does the pipeline of future projects include a variety across the enterprise (some from each functional area)?	X	
Are certified belts being redeployed on other initiatives?		X
Change Management	**Yes**	**No**
Have recent project successes been communicated across the enterprise?	X	
Has a repository of lessons learned and best practices been established?	X	
Has a reward or recognition program been deployed for LSS team members?	X	

The keys areas for any LSS audit include training, project selection and completion, change management, and sustainment. I typically rank an organization "L" or low in maturity if they can only answer yes to 10 or fewer of the items, "M" or medium if they can answer yes to 11-15 of the items, and "H" or high if they can answer yes to more than 16 of the 19 questions.

In this example, we realize that the overall maturity of the current organization is "M" (since there were 14 yes responses), and that the key areas for improvement are focused on training and the redeployment of certified belts. This is often the case in the public sector, where most certified belts are rarely immediately redeployed on a new project.

Affinity Diagrams

Impact in Overcoming Public Sector Challenges								
Breaking down stovepipes	Creating urgency	Leadership support	Metrics	Common goals	Increasing customer focus	Reducing impact of turnover	Overcoming complexity	Fostering collaboration
				X			X	X

As you begin to identify a large number of potential improvement initiatives, it may become apparent that some redundancy exists. This is where the power of affinity diagrams can help to focus future initiatives into key areas. (See Figure 3.4.) The purpose of affinity diagramming is to categorize ideas as they become apparent (that is, identify affinities in the information available). For example, if you identified call wait time, speed of closing incoming calls, and resolving customer service issues, these all may be categorized as call center opportunities. This information can be used to focus a round of improvement efforts on the call center.

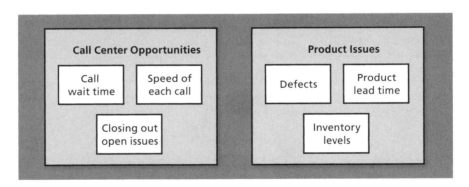

Figure 3.4 Example affinity diagram.

The easiest approach to affinity diagramming is to first list all possible improvement ideas on separate sheets of paper. Then, ask the same team members who developed the ideas to group them into categories based on similarities. Affinity diagramming helps identify initial high-value areas for improvement and it also may reduce the resources required to develop, implement, and sustain improvements by eliminating redundant project charters.

Multi-voting

Impact in Overcoming Public Sector Challenges								
Breaking down stovepipes	Creating urgency	Leadership support	Metrics	Common goals	Increasing customer focus	Reducing impact of turnover	Overcoming complexity	Fostering collaboration
X								X

So your LSS project team has reduced a list of possible high-value initiatives to three or four, but you cannot reach a consensus on the most impactful project. A tool that helps a team reach consensus in a non-combative manner is the multi-voting approach. The steps are outlined here:

1. Determine how many project or improvement ideas you would like to include in the vote (for example, 7 ideas).
2. Multiply that number by 2 to arrive at a total number of votes (in this example, 14 votes).
3. Ask individual team members to allocate these votes to the various projects as they prefer (for example, they can allocate all 14 votes to one idea or split them up as desired).
4. Once each team member has completed a silent vote, tabulate the results and rank the ideas according to number of votes cast.
5. Move forward with further defining the topic that received the highest number of votes. Document the outcome of your entire multi-vote session to consider in future discussions.

Note: Multi-voting can be used whenever your project team is trying to reach consensus.

PICK Diagram

Impact in Overcoming Public Sector Challenges							
Metrics	Common goals	Leadership support	Breaking down stovepipes	Creating urgency	Overcoming complexity	Fostering collaborative environment	Reducing impact of turnover
	X	X	X	X			X

Now that you have a stack of project charters you'll have to decide which project should be completed first. A PICK diagram is a good tool to use to visually depict the highest-value initiatives. Based on the information included in your project charter (goal statements, business case, opportunity statement), place each potential project in the matrix illustrated in Table 3.14.

Table 3.14 PICK diagram categories.

Category	Ease	Costs	Benefits
Possible	Low	Low	Low
Implement	Low	Low	High
Challenge	High	High	High
Kill	High	High	Low

Possible – These initiatives would be expected to have a low to minimal impact on the overall organization, but are easily implemented at a low cost. Additional analysis may be required to determine whether the benefits outweigh the overall costs.

Implement – These initiatives would have a high impact on the organization; they are easily implemented at a low cost. Projects that fall into this category should be implemented as soon as possible.

Challenge – These initiatives may have a significant impact on the organization, but may be difficult to implement or come at a high cost. Initiatives that fall into this category should be investigated further to determine whether the benefits outweigh the costs.

Kill – These initiatives would have little or no impact on the organization; they are difficult or costly to implement. Improvements that fall into this category should be eliminated from the list of possible recommendations for implementation.

A PICK diagram template is illustrated in Figure 3.5.

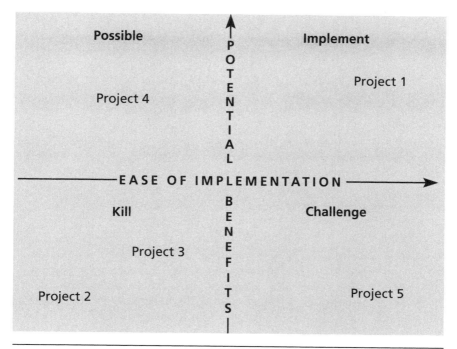

Figure 3.5 Example PICK diagram template.

In this example, the recommendation to leadership is apparent: immediately move forward to implement Project 1 (benefits and ease to implement are high at a low cost); investigate Project 4 in further detail (it has a high potential benefit but may be difficult and costly to implement); investigate Project 5 in further detail (it may be easy to implement, but the benefits may be low); and eliminate Projects 2 and 3 (the payoff is likely low, while the costs are high).

WHY LEAN-SIX SIGMA PROGRAMS FAIL

1. Lack of leadership commitment. The number-one reason LSS programs fail is a lack of leadership commitment. This can be overcome by clearly articulating expectations, communicating project successes, and showing the overall impact on leadership's key performance measures (productivity, customer satisfaction ratings, profit).
2. Organizational fear of change. Fear of change can have a significant impact on your success. A proper change management strategy and the integration of all levels of an organization in the deployment will mitigate this risk.

3. Egos. Egos can play a big role, especially if the person who built the product or process 30 years ago is still part of the organization. This is another reason to make sure you not only have an executive steering committee, but also continuously work to gain buy-in from every stakeholder. Another easy way to overcome egos is to involve the affected stakeholder in the *entire* improvement process.
4. Thinking "it won't work here." This is where quick, sustainable, consistent results help gain and maintain support.
5. Failure to implement or implementation phobia. This is where risk management plays a critical role. The possible downside of implementing an initiative must be clearly articulated. If the benefit will outweigh the possible detrimental effects 90% of the time, there is no reason not to try.
6. Lack of proper training, coaching, and mentoring. It can't be a "set it and forget it" approach to training. Training must be part of your human resource planning (these will be highly valued full-time resources). It may seem costly, but when you factor in that a Black Belt could save you $1 million/year, you'll see that it is a small price to pay. You also need to think about career development for these resources (BBs becoming MBBs, MBBs becoming Operation Directors).
7. Using an all-or-nothing approach to the tools. *Do not* try to use each and every tool on every project. Use whichever tools will give you sustainable results the quickest.
8. Lack of an enterprise-wide view for your LSS program. This is another place where the involvement of all stakeholders will assist in breaking down organizational stovepipes.
9. Unrealistic initial goals. It would be a significant stretch goal to save $100M on the first project. Projects like this are commonly referred to as "trying to boil the ocean." LSS is focused on improving the enterprise through "bite-size chunks," with a goal of continuous process improvement.
10. Using LSS when it is not the right fit. LSS is not the only valuable methodology in the improvement toolkit. If it would be more beneficial to use ISO, TOC, or another methodology, it is the responsibility of the LSS leader (BB or MBB) to revamp the project charter appropriately.

11. Using the wrong tools. It is essential to ensure that simulations are tailored to your environment. If you use a manufacturing simulation in a transactional environment, you may lose buy-in from you stakeholders.

Public Sector Project Highlight

Self-Sustaining Improvement Program

The goal of any LSS program is to be self-sustaining from a funding perspective. In the public sector, lead time for new funding can be significant (a year or more for most congressional appropriations). One way to overcome a lack of funding is to create business cases that provide cost savings significantly larger than the resources required. This will often lead to funding for a specific initiative (which will be re-infused using the funding re-allocation plan) and provide resources to facilitate other high-value LSS initiatives.

SPOTLIGHT – BUILDING EFFECTIVE TEAMS

The project team can make or break any LSS initiative. Regardless of the opportunity, if the team is not working at optimal levels the project will likely be sub-optimized or, even worse, a complete failure.

The first step in building an effective team is to identify which key stakeholders must be involved. For details on how to approach this, please refer to the Integrated Stakeholder and Communication Plan section.

The first step is to establish ground rules such as these:

- Meeting dates and times will be approved by leadership.
- Meetings will be held in a non-attribute environment.
- Everyone will have a voice at the table and an opportunity to use it.
- Distractions will be kept to a minimum (laptops and cell phones will be turned off).
- Preferred communication methods will be agreed upon (e-mail or phone).
- Respect will be provided at all times (this includes showing up on time with assignments completed).

Once the team is identified and ground rules are discussed, the next step in the process is to identify, assign, and agree upon clear roles and responsibilities. The last step is often overlooked. Even if someone is not looking forward to having a role on the project, at this point we must at least ensure there is acceptance of the role. One way to overcome roles and responsibilities issues is to reiterate WIIFT (what's in it for them). This involves discussing how the project or possible improvements will affect them personally:

- Career development
- Exposure to other areas of the organization
- More effective use of time
- More efficient processes
- Reduced frustration
- Increased personal visibility

Now that the team has been identified, roles and responsibilities have been agreed upon, and there is a solid commitment secured from leadership, you can move forward with your project.

This is not to say that the typical forming, storming, norming, performing, and adjourning will not occur. However, once a common goal is agreed upon and clear roles and responsibilities are established, it will be easy to overcome any issues. Remember to keep the team focused on the leadership-approved project charter. This will help later when improvements are being implemented, since all levels of an organization will have been involved in one aspect of the initiative or another.

Section 4
Focus on Waste, Then Variation

"When you're finished changing, you're finished."

– BENJAMIN FRANKLIN

Why waste time standardizing a broken process? The initial focus should be on lean concepts when an organization uses the LSS toolkit for the first time. Lean tools typically provide significant reductions in waste (rework, cycle time, and so on) and implementation of lean improvements is normally faster than when using Six Sigma concepts. Using lean first will help you assure the public sector organization obtains quick results and maintains leadership support.

VALUE OF A PROJECT KICKOFF MEETING

Impact in Overcoming Public Sector Challenges								
Breaking down stovepipes	Creating urgency	Leadership support	Metrics	Common goals	Increasing customer focus	Reducing impact of turnover	Overcoming complexity	Fostering collaboration
X	X	X		X	X			X

Prior to starting any LSS initiative, it's important to ensure that everyone is playing from the same sheet of music. In terms of a LSS initiative, this means that each team member understands the high-level concepts of LSS (what is LSS, why is it being used, what outcomes do we hope it will provide), understands the roles and responsibilities involved from a macro-level (Black Belt, Project Sponsor), allows the Project Sponsor to clearly show support for the initiative, and provides an opportunity to network informally among the project team members. A project kickoff meeting should be held in advance of each LSS improvement initiative and should involve as many identified key stakeholders as possible (keeping in mind that some key stakeholders may not have been identified at this point). These meetings require minimal effort and only about an hour of everyone's time, but they will help set the stage for a successful project and support a sense of teaming from the outset. See Table 4.1 for an example of a project kickoff meeting agenda.

Table 4.1 Sample project kickoff agenda.

- Champion kickoff
- Project charter review
- Project team identification
- Identification of training requirements
- High-level resource requirements discussion

PROJECT CHARTER – ONE-STOP SHOP FOR WHY THE PROJECT IS BEING PURSUED

Impact in Overcoming Public Sector Challenges								
Breaking down stovepipes	Creating urgency	Leadership support	Metrics	Common goals	Increasing customer focus	Reducing impact of turnover	Overcoming complexity	Fostering collaboration
X	X	X	X	X	X	X	X	X

The project charter is one of the most critical items for any LSS initiative. It includes both the rationale and the authority to complete a project. When created appropriately, the project charter becomes the foundation for maintaining leadership and stakeholder support. It will assist in keeping the project tightly scoped and ensure that your improvements are supporting the initial goals of your approved charter.

The first step in creating a project charter is the creation of an opportunity or problem statement. This statement should clearly articulate why the project is being pursued. The purpose of this statement is to set the stage for discussing goals, stakeholder involvement, and project scope. All of these items, as shown in the project charter example in Figure 4.3 later in this chapter, will be based on this initial opportunity statement.

Having identified the initial opportunity statement, we now must begin to involve all relevant stakeholders. A solid LSS team consists of a LSS team lead (Master Black Belt, Black Belt, or Green Belt), a project champion, a project team lead (this should be the process owner, if at all possible), and functional subject matter experts. The key to having a robust team is the integration of stakeholders across the entire value stream, from suppliers to the end customer. It may not be easy to determine where the variation or waste is occurring unless all relevant stakeholders are present.

Business Case

Once you have identified a potential opportunity and the stakeholders involved, you must validate that the potential results will be more than the resources required (that is, potential benefits exceed forecasted costs). The data to complete this analysis should be provided directly from your stakeholders and should be consistent with a rough order of magnitude estimate. If data is available, a quick Pareto chart (discussed later in this book) can provide potential projects and clearly display the potential impact. Keep in mind that there will be variation in your final outcome. At this point, the goal is to clearly display to leadership that this project will provide value.

High Level Description of Current Process

With your identified stakeholders, discuss the high-level process related to your opportunity. The goal is to be able to articulate the process in one to two sentences while also identifying potential parallel processes. These are important because as you complete your LSS project you may find that you are improving the wrong process. When this happens, the most probable cause of waste or variation is a parallel process. When you implement standard operating procedures or training updates, you must ensure that any training that references your process is updated as well.

SMART Goals and Objectives

Before starting any improvement initiative, you must be able to justify a quantitative benefit to the organization or the project may not be approved from a cost vs. benefit perspective. Establishing initial goals and objectives, also known as your initial improvement targets, will assist senior leadership in visualizing the impact of your initiative on the enterprise and its bottom line. For your specific project, these must be directly linked to a goal, objective, or metric in your strategic plan. If your project charter goal or objective does not clearly link to one of the items from your strategic plan and there is quantifiable value in completing the initiative, it may be necessary to revamp your strategic goals, objectives, or balanced scorecard.

Goal vs. Objective Statement

The terms *goals* and *objectives* are often used interchangeably and incorrectly. A goal in terms of LSS is your overarching improvement target. Objectives are the manner in which those goals are to be achieved. For example, a goal may be to increase profitability by 10% by the end of the second quarter 2012. Related objectives might be to decrease costs by 15% in current product lines by the first quarter of 2012, increase the customer base by 10% by the third quarter of 2011, and so on. As depicted in this example, the goal statement is the overarching accomplishment an organization strives for, whereas the objective statements are how the organization plans to achieve the goals.

Quantitative vs. Qualitative

Goals and objectives should incorporate a quantifiable target. Quantitative goals and objectives have specific number targets involved, whereas qualitative goals and objectives do not. For example, decreasing costs by 15% is a quantitative goal; increasing customer loyalty is a qualitative goal. In another example, customer feedback may state that the pizza being prepared in the food court has too much sauce. In order to provide a quantitative improvement, more information would be required. If a survey were facilitated, feedback might show that 95% of customers surveyed would like 8 ounces of cheese and 4 ounces of sauce. The quantitative improvement metric for this example is a perfect ratio of 2 to 1 for cheese and sauce in order to satisfy 95% of the customer base. It is more constructive to use quantitative goals and objectives as part of your improvement initiatives, as it is easier to identify specific measurements in the Measure phase that can impact your quantitative goal and objective.

Project Scope

Project scope is one of the most critical aspects of your project charter. If it is not accurately defined with all key stakeholders at the beginning of your project, you are likely to have significant scope creep. Scope creep occurs when additional requirements, goals, or objectives are added to your project and therefore expected as outcomes. A good tool to use to prevent this from occurring is the in-and-out-of-scope tool. See Figure 4.1.

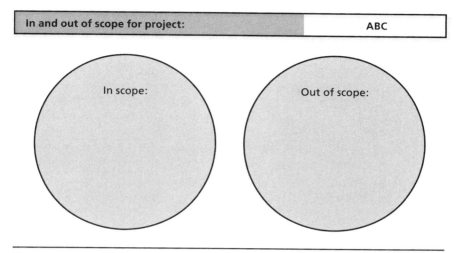

Figure 4.1 In-and-out-of-scope tool.

Beyond identifying what is in scope and what is out of scope, you will also establish the start and finish dates for your project using two process steps. One is the "trigger" or start of your process and the other is the "point of termination," the end of your process. It is beneficial to list parallel processes as part of scoping an effort. It may become apparent that it is not the original process that is causing variation or waste, but rather a parallel process.

High-Level Milestones

From a milestones perspective, important information to include in the project charter is phase completion dates. This will provide leadership and all team members with an understanding of when the project will be completed and the amount of time provided for each phase. As a secondary effect, this will enable you to keep your LSS team motivated, as you have provided leadership with a tentative schedule of activities.

Example Project Charter

Completing these tasks results in an overall project charter that outlines the key information for your initiative. See Table 4.2 for a recommended format for displaying all this information in a concise manner so as to obtain leadership approval.

Table 4.2 Example project charter.

Project name	Project ABC
High-level description of current process	Process is utilized to order supplies for the entire XYZ department.
Project champion	Mrs. Jane Doe
Project owner	Mr. John Smith
Project facilitator	Bill Jones, LSS BB
Business case	Reducing the lead time for supplies would reduce the amount of supply inventory required. As a secondary effect, the need for employees to hoard supplies that run low would be eliminated. If the goal is attained, it will reduce supply inventory by 40%.
Opportunity statement	Extensive time required for supply replenishment process is causing several stakeholders to double safety stocks. This is tying up 20% of the budget in unnecessary supply inventory.
Scope	Start: Supply requests received by John Smith. Finish: Supplies unloaded in supply cabinet by John Smith.
Parallel processes	Supply expense reporting
Goal statement	Decrease the supply replenishment lead time from 17 days to 7 days by the end of the 2nd quarter.
Potentially affected organizations	Procurement, accounting, and delivery departments
High-level milestones	Define — 01/01/XX; Measure — 02/15/XX; Analyze — 04/15/XX; Improve — 05/15/XX; Control — 06/15/XX

SIPOC

Impact in Overcoming Public Sector Challenges									
Breaking down stovepipes	Creating urgency	Leadership support	Metrics	Common goals	Increasing customer focus	Reducing impact of turnover	Overcoming complexity	Fostering collaboration
X					X		X	X

The SIPOC, which stands for suppliers, inputs, process, outputs, and customers, is the first glimpse into the process and acts as an initial high-level process map. The purpose of this tool is to build upon the data gathering that has been done to this point in the Define phase and to prepare for potential process mapping efforts. It provides fundamental information on who supplies inputs to the process and who are the customers that receive the process outputs.

There is a quick and easy approach to completing a SIPOC:

1. Using the SIPOC template (see Figure 4.2) and the completed project charter, fill in the start or trigger for the process, the last step or end of your process, and any specific items that are not included as part of the project scope.
2. Brainstorm the outputs of your process.
3. Based on the outputs listed, identify the customers who receive those outputs.
4. Next, brainstorm the inputs for each output.
5. Finally, identify the suppliers of the inputs you have listed.

Suppliers	Inputs	Process	Outputs	Customers
		Start:		
		①		
		End:		
⑤	④	①	②	③
		Does not include:		
		①		

Figure 4.2 SIPOC template.

As you can see once again, the tools continue to build off each other. This one uses a significant amount of the information from your project charter. It is ideal to complete these types of exercises with your core LSS project team. This helps to begin laying the groundwork for the change management plan and overall project success.

Completing a SIPOC seems like a simple exercise, but it will begin to lay the fundamental groundwork for discussing the entire value stream from suppliers to ultimate customers. It will assist in eliminating waste and support the creation of a sound measurement plan by identifying all inputs and outputs of the process as part of the LSS initiative.

VO"X"

Impact in Overcoming Public Sector Challenges								
Breaking down stovepipes	Creating urgency	Leadership support	Metrics	Common goals	Increasing customer focus	Reducing impact of turnover	Overcoming complexity	Fostering collaboration
	X		X	X	X			X

As part of your continuous process improvement efforts, it is important to continuously seek out the voice of your customers. Keep this in mind when you begin to contact your customer base. If you can build a strong, collaborative relationship with them on the front end and strive for a win-win relationship, gathering their insights in the future will be greatly simplified.

Beyond the voice of the customer, there are additional "voices" that must be considered from an enterprise-wide value stream perspective: the voice of the business, the voice of the supplier, the voice of the administration, and the voice of the employee. Remember, LSS encourages the elimination of waste throughout the entire value chain.

The intent is to foster a non-attribute environment where everyone feels comfortable providing open and candid feedback. This will provide the framework for taking seemingly qualitative themes and translating them into quantitative requirements, metrics, and improvements. This, in turn, provides the data and information necessary to drive customer "feelings" into business and technical requirements.

Types

Many voices impact the overall success of an organization including customers, suppliers, employees, the current administration, leadership, and a multitude of others. It's important to identify the internal/external stakeholders who have an impact on the ability of an organization to deliver the mission, vision, goals, and objectives as part of a strategic plan. The voice of these stakeholders can then be further quantified into critical-to-quality elements or CTQs. Although this should not be considered an all-inclusive list, some of the more frequently used Vo"X" areas are discussed here.

Voice of the Administration (VoA)

This includes the strategic vision, mission, goals, and objectives of the current governmental administration (for example, the current governor or president or congress). Although they may not always directly impact the charter of specific public sector entities, an organization must be aware of their focus areas so that their respective strategic plans can be tied to them appropriately. The current administration will likely impact funding availability, which will have an impact on the outcome an organization can achieve.

Voice of the Employee (VoE)

Employees are often overlooked when it comes to the Vo"X" process. This is likely due to the internal role they play. Employees will be involved as part of the initial strategic planning process, and it's important to identify detailed linkages for keeping them engaged. Some potential areas from a VoE perspective include employee retention, education, and financial incentives (salaries and bonuses). If the employees voice is not integrated as part of the organizational CTQs, it is likely the goals and objectives will suffer.

Voice of the Supplier (VoS)

Internal and external suppliers of information, services, and products are commonly overlooked in the public sector. Although an organization may be required to provide certain inputs, those inputs may not be provided in the most efficient and effective manner possible. Therefore, obtaining the VoS is critical to optimizing the enterprise and should be appropriately integrated into the final CTQ.

Voice of the Customer (VoC)

The voice of the customer is the most commonly used part of Vo"X" approach. It involves engaging the end user to ensure an organization is providing appropriate products or services:

- What is truly required (only things that add value; no waste)
- When it is required (for example, the first of the month)
- At an appropriate interval (for example, monthly, weekly, or daily)
- In an appropriate format (for example, MS Excel format)
- To the appropriate person/location (for example, to accounting or logistics)

Customer Segmentation

Before we discuss the process for customer segmentation, we must discuss the value it provides. The basic value of this approach to Vo"X" provides two significant benefits. The first is that it identifies which customers have the most impact on your organization. From a public sector standpoint, this is typically defined as who provides you with the most funding or who your largest customer group is. This can overlap with who takes up the majority of your organization's time, but if it does not, that should be identified as an opportunity for improvement. (If a customer takes up 75% of your total time available but only makes up 5% of the total customer population, there is likely an opportunity for improvement to reduce the amount of time spent on such a small stakeholder group.)

The other key benefit of customer segmentation is that the results can help determine a valid statistical sample. As part of any sample, there should be adequate randomness to the sample selection, including all possible customer groups. Samples (a group of selected items from a population, such as 1,000 U.S. citizens) are often used throughout LSS in place of populations (all possible items, such as all U.S. citizens). Although samples do not include all data possible, using samples typically saves time, money, and other resources because the data is easier to collect. See Table 4.3.

Table 4.3 Example of customer segmentation.

Senior executive service employees	5%
GS-12 — GS-15 employees	25%
GS-9 — GS-11 employees	35%
GS-8 and below employees	35%

One way to identify potential customers and segment them into groups is to first complete a SIPOC for the respective process. This will provide all possible customers for a particular initiative. The customers can be left as initially grouped or, if there is not enough granularity in the initial customer segmentation to identify a valid sample, they can be further stratified (split into more descriptive categories). As an example, consider the United States Congress. If this does not provide enough granularity for your Vo"X" exercise, you could split the segment further along party lines (Democrats and Republicans) or you could split it according to years of experience (1-10, 11-20, and 20 or more). It is up to you and the project team to determine an appropriate customer segment.

Once you have determined a list of customer segments and collected data relative to their size/scope (funding, number of customer requests, and so on). This will help to ensure you are driving specific customer segments to true critical-to-quality items as discussed later in this section.

Gathering Vo"X"

There are many ways to gather data respective to your external stakeholders (customers, suppliers, and so on). Some of the primary methods include interviews, focus groups, and surveys. These methods should not be used interchangeably, as they provide a varied amount of detail and come with disparate costs in terms of time, effort, and other resources. Figure 4.3 provides a high-level overview of when each type should be used.

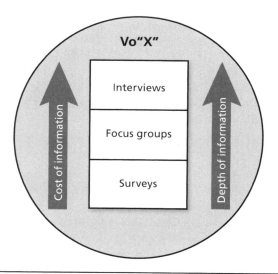

Figure 4.3 Relationship of Vo"X" cost vs. depth of information.

Interviews

Interviews are a common Vo"X" technique in which interviewers converse with interviewees via web-chats, phone, web-cam, or in person. They can occur in a multitude of formats including structured, non-structured, and hybrid approaches. The format and delivery method should be tailored according to the type of information desired. (See Table 4.4.)

Table 4.4 Example interview agenda.

- Provide overview of LSS project
- Review interview ground rules
- Provide goals of the interview
- Ask detailed interview questions
    - What are your thoughts on our current training program?
    - How long have you been with the organization?
    - What do you believe are some of the key opportunities for improvement?
- Provide overview for next steps

Before talking to any of your customer groups, define the interview parameters:

- *What are your goals?* If everything were to go exactly as planned, what type of information would be gained from an interviewee? This can include their thoughts on the process, what they define as value, or any other information that may provide value as part of your LSS initiative. If you are unable to identify the goals of the interview, it is likely that the initiative was not tightly defined or scoped as part of the project charter.
- *What resources are available to facilitate the interview?* If you have limited funding for travel, it is likely that a web- or phone-based interview will be appropriate. If time is limited, you may prefer to use a different Vo"X" approach such as a focus group or survey.
- *What type of interview format will you use?* It is almost always better to use either a structured or hybrid approach for LSS initiatives because this will ensure similar inputs as part of each interview. Using open-ended questions as part of a hybrid approach can solicit additional insights about what the customer perceives as value, improvement ideas the customer may have, or general thoughts on the overall process.

The information gained from an interview is typically very detailed, but it can be skewed toward a specific customer experience. If resources are available, it's desirable to facilitate a significant number of interviews as part of your Vo"X" strategy. If not, other approaches may provide for more significant returns.

Focus Groups

Focus groups are similar to interviews in that you will be interacting directly with your stakeholders. The primary difference is that during the focus group there will be eight to twelve people providing feedback instead of just one or two. This can increase the amount of varied feedback obtained, but it will likely reduce the depth of information obtained and may increase the potential for "group think" (people tend to be less open about their personal feelings and prefer to go along with the majority of the group). This is not to say that focus groups are not valuable sources of information; just remember to take these caveats into account when interpreting the data.

As with an interview, identify the goals, resources available, and format in advance of the focus group. Have these components reviewed and approved by the project sponsor. As a caveat, I have seen the format of the focus group directly impact the amount of data obtained (whether it is face-to-face, via videoconference, or over the Internet).

Surveys

Although surveys may not provide the depth of information that an interview or focus group can provide, they are often much less costly and typically provide more quantitative data. Surveys help ensure a standardized response format, can be either rigid (multiple choice, yes/no) or open (fill in the blank) in terms of responses, and can follow a variety of delivery approaches (phone, web-based, e-mail). A typical survey should take no more than about 15-30 minutes to complete or respondents may become frustrated and abandon the effort.

This brings me to my next point. What is involved in a robust survey? (See Figure 4.4.)

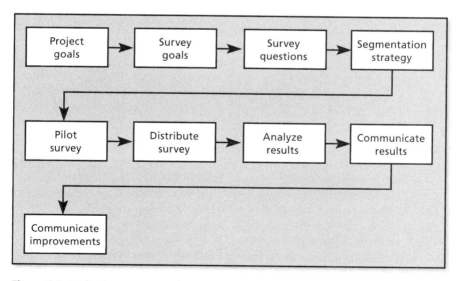

Figure 4.4 A robust survey approach.

1. Project Goals – The goals for the survey should be clearly aligned to the mission, vision, goals, objectives, and scorecard for the organization. As part of the survey, these goals should be clearly articulated to the survey respondents.

2. Survey Goals – These are the goals for the specific survey questions. For example, the project goals may focus on increasing customer satisfaction. As part of a specific survey, you may be trying to better quantify satisfaction.

3. Survey Questions – When the goals of the survey have been identified, you can start to identify specific questions. Following the above example on measuring customer satisfaction, one question might be, "On a scale of 1-5, how does the timeliness of information for Organization X affect your overall satisfaction with the service?"

4. Segmentation Strategy – Did you ever wonder why surveys ask for your income levels, age, and occupation? This information allows researchers to accurately segment that data once it has been collected. The segmentation of data provides significant value as you are analyzing it (for example, Pareto's, histograms) by identifying key improvement areas. Specific segmentation information can include location, customer type, timing, and more.

5. Survey Pilot – Before any survey is distributed to a large stakeholder group, it must be piloted. A survey pilot should be randomly distributed across the various survey target areas. Once the random participants have completed the surveys, the team should identify any areas for improvement in terms of the information obtained and ensure that it meets the survey goals and that it can be easily analyzed. If there are areas of concern, the survey should be changed to overcome those concerns or the survey team should reach back out to the survey respondent to identify why the issue occurred. A survey pilot will not only ensure that you are gaining the information required, but also help avoid having to go back out to the survey respondents at a later date.

6. Survey Distribution – Once you have piloted the survey and given it the "green light," distribute it across all stakeholder groups (either as a stratified sample or to the entire population based on resources available). Ensure that the method of retrieval is optimal for your proposed analysis (for example, spreadsheet, e-mail) and that a point of contact is identified to handle any questions or concerns that occur.

7. Analysis – Once all of the data has been collected, the results of the survey can be analyzed using Pareto charts, histograms, scatter diagrams, and other statistical tools.

8. Communication of Results – The results of the survey should be communicated to leadership, the project sponsors, the project team, and to all survey respondents if possible. This helps further foster an environment of open communication.

9. Communication of Improvements – Once survey results have been used to implement sustainable improvements (using the other LSS tools discussed in this book), the improvements made should be communicated back to all survey respondents as part of the overarching communication strategy. This helps demonstrate that their 15-30 minutes of effort had a significant impact on the organization.

Using a robust survey approach helps ensure a linkage to the project and organizational goals. Communicating these linkages helps make the buy-in and change management process much easier. Through surveys and other Vo"X" tools, you can not only obtain the information necessary to make an impact, but also continue to gain buy-in from all stakeholders.

Critical-to-Quality (CTQ) Elements

Impact in Overcoming Public Sector Challenges									
Breaking down stovepipes	Creating urgency	Leadership support	Metrics	Common goals	Increasing customer focus	Reducing impact of turnover	Overcoming complexity	Fostering collaboration
X	X		X	X	X		X	X

Now that you have listened to the voice of your customers, suppliers, business, the administration, and everyone else impacted by the product and service offered, it is time to link that feedback to critical elements. Critical-to-quality elements, which we will use as an all-inclusive term (critical to the business, critical to the administration, critical to suppliers, and so on), quantify the Vo"X." Unfortunately, the initial Vo"X" is normally qualitative information. The purpose of CTQs is to drive that qualitative information obtained through interviews, focus groups, and surveys into quantitative measures of success.

As an example, let's use the feedback obtained during discussions with your customers related to customer satisfaction. The feedback showed that timeliness and defects were the primary components of satisfaction. Therefore some of your CTQ relative to customer satisfaction may be delivery within two days of order initiation and less than one defect per 1,000 customer orders. Once you have worked through all of the items identified as part of your Vo"X" exercise and identified relative CTQs, you can easily plot them on a CTQ tree. (See Figure 4.5.)

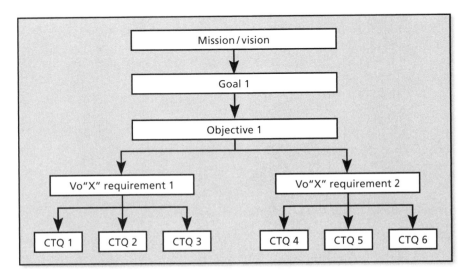

Figure 4.5 Example CTQ tree.

The importance here is making sure that the CTQ elements, similar to your goals and metrics, are linked to the strategic goals and objectives of the organization. If they are not, then either the customer requirements are not part of your core mission and vision or it may be necessary to revamp the strategic plan.

> **Public Sector Project Highlight**
>
> **Is That Really a CTQ?**
>
> As people begin to use the LSS tools for the first time in the public sector, there is sometimes a fear of pushing their LSS team members, subject matter experts, and especially leadership too far. This fear is enhanced in the "rank and file" environment of most public sector organizations. Although this fear is valid, it must be overcome because of the critical importance associated with driving to true customer requirements. For example, if someone states that an increase in communication is required, the LSS or project lead must drive this into a "real" requirement. This could mean an increase in meetings or status updates or simply a phone call when something has been completed. If CTQ elements are not driven down into tangible requirements, it will be impossible for the team to focus in on improvements. Project sponsors, executive steering committee members, and the project team lead should provide significant support for "pushing" personnel to provide what is required.

Kano Analysis

Originally developed by Noriaki Kano, a Kano analysis helps determine how certain requirements will either satisfy customer needs, delight customers, or dissatisfy a need. Completing this analysis is helpful in determining areas of focus (anywhere dissatisfaction or satisfaction needs exist) while also visually displaying areas where customers' expectations could be exceeded. Figure 4.6 illustrates a Kano model and its impact on customer satisfaction.

So what does this mean? The easiest way to understand Kano analysis is by completing an example. People who stay in hotels have certain expectations: a bed, a bathroom, towels, running water, and a television. These are classified as satisfiers when present (ready and available in the hotel room) and dissatisfiers when they are missing (too few towels or a television that does not work). A delighter in this case may be 1,000 thread-count sheets or a cookie placed on the pillow each night. See Table 4.5 for a Kano analysis output table for this example.

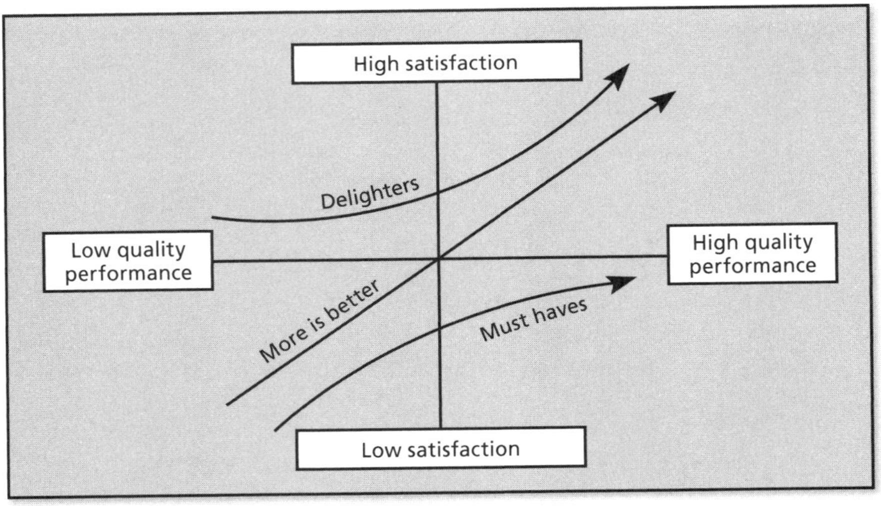

Figure 4.6 Kano model.

Table 4.5 Example Kano analysis output table.

Example Kano output – hotel room	
Must haves	• Bed • Bathroom
More is better	• Television options • Free Internet
Delighters	• Warm cookies • Bed turn-down service

In this example, the true requirements are a bed and bathroom. The more television options and free Internet available, the happier the customer (more is better). Although not expected as a normal part of the customer experience, warm cookies and a bed turn-down service would delight the customer. These items can be identified as part of the Vo"X" collection and used to develop a process, product, or service that continuously exceeds customer expectations and helps support increased funding and customer satisfaction.

Integrated VO"X" Approach

Even though there is a continued focus on the Vo"X" as part of the LSS methodology, a public sector organization should focus on improving the overall enterprise. With this in mind, each "voice" should also be tied to a relative organizational impact. For example, the customer may have voiced that a critical factor in their loyalty is a maintenance repair within 10 days of notification (this is your Vo"X"– CTQ element). It must be appropriately tied to an organization's mission, goals, or objectives. (See Figure 4.7.)

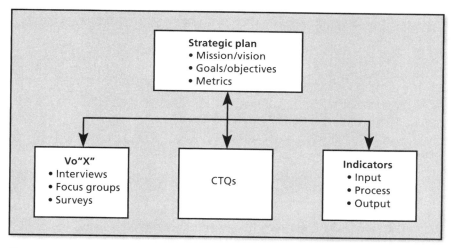

Figure 4.7 Integrated Vo"X" approach.

If an organization is unable to tie it to one of these items, then it may not add value overall. Maybe maintenance turn-around time is not the organization's focus, but rather accuracy of maintenance or ensuring compliance with certain specifications. In this example, although it may add value to deliver within 10 days as a customer delighter, it is really not part of the organization's charter.

> **Public Sector Project Highlight**
>
> **Vo"X" = The Ultimate Customer Experience**
>
> Vo"X" was used on a recent enterprise-wide LSS initiative focused on the development of an IT solution that would help capture budgetary requirements at the lowest levels of an organization and provide a rolled-up picture for an entire appropriation. This effort included using voice of the customer tools such as customer segmentation, interviews, surveys, CTQ's, and Kano analysis to define all functional requirements. These functional requirements were then driven into a detailed technical design document focusing on providing the "ultimate" customer experience while obtaining the required information. The results of this initiative were successfully implemented across a global enterprise. This initiative also eliminated an obsolete process, saving the organization thousands of work hours annually.

BASELINE ENTERPRISE-WIDE PROCESS MAPS

Impact in Overcoming Public Sector Challenges								
Breaking down stovepipes	Creating urgency	Leadership support	Metrics	Common goals	Increasing customer focus	Reducing impact of turnover	Overcoming complexity	Fostering collaboration
X	X	X	X	X	X	X	X	X

When you begin to use LSS in a public sector organization, the normal questions come to mind. What should we focus on for our initial projects? How will we begin to gain leadership buy-in? What do we use as our baseline for continuous process improvement? One tool in the LSS toolkit can provide answers to these questions and more. EWPM, or enterprise-wide process mapping, can assist in fundamentally changing an organization through the identification of key process indicators (used for defining your initial goals and objectives), eliminating organizational stovepipes, identifying high-value initiatives to meet your goals, laying the foundation for standardized operating procedures across the entire enterprise, fostering an environment of continuous process improvement, providing organizational continuity, and gaining leadership buy-in through initial project successes. From the use of one tool you can lay the fundamental foundation to ensure LSS is a success in a public sector organization.

One of the most significant impacts EWPM can have on a public sector organization is creating a collaborative environment or, at a minimum, opening communication channels between functional organizations. In creating the initial EWPMs, you will be required to create cross-functional teams specific to a given process. These teams will map out the process and identify any independencies between organizations. (See Figure 4.8.)

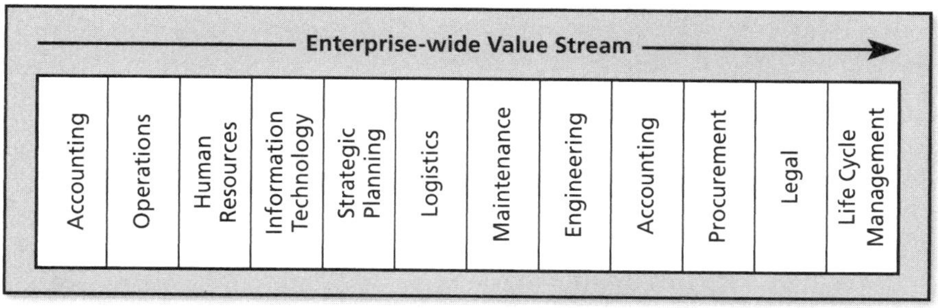

Enterprise-wide Value Stream
Accounting	Operations	Human Resources	Information Technology	Strategic Planning	Logistics	Maintenance	Engineering	Accounting	Procurement	Legal	Life Cycle Management

Figure 4.8 Example enterprise-wide value stream.

Have you ever heard this from other functional organizations? "That's the way we have always done it here" or "I don't know what they use it for, but accounting needs that report every month." Never underestimate the value of bringing key stakeholders together to map out processes or discuss improvement opportunities. As a secondary affect, you may also be able to gain insight into some potential "ground fruit." Some examples of ground fruit, also known as high-value quick wins because they are low-cost and high-yield, could include identification of reporting requirements that no longer add value, process hand-offs that do not add value, or opportunities to eliminate or streamline sub-processes.

> **Public Sector Project Highlight**
>
> **EWPM**
>
> In working to transform a large public sector agency's processes, a LSS team identified the planning process for the next fiscal year as a high-value initiative (HVI). The project was initially scoped as a LSS DMAIC initiative. As the team worked through the Define phase of the project, it became increasingly evident, through obtaining the voice of the customer, that the process was severely outdated, cumbersome, and of little value to customers. With that in mind, the team re-scoped the initiative as a LSS DMADV (Design for Six Sigma) project and was able to create a streamlined process to accurately and clearly define an organization's funding requirements to include what the funds will provide in terms of mission capabilities. While providing more value to the customer, we were also able to eliminate non-value-added information, which equaled a 90% reduction. We drastically reduced the burden in producing the previous product, refocusing the available hours on analysis instead of non-value-added data gathering.

One of the first steps in every LSS deployment should be a baseline assessment of all existing processes. You can use this baseline to identify your enterprise-wide improvement goals. EWPM provides a unique systematic approach to mapping out each of your organization's key processes and quantitatively assessing key metrics within each process step (processing time, lead time, and so on). This will ultimately allow you the opportunity to create a cumulative baseline for your organization (see Figure 4.9).

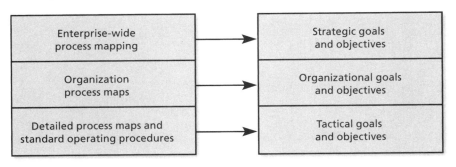

Figure 4.9 Granularity of process maps required to achieve respective goals.

Once you have a fundamentally sound understanding of your current state or baseline through the EWPM effort, you can then create improvement metrics. As part of gaining leadership support from the beginning of your deployment, the appropriate measures should be taken to guarantee senior leadership agrees with your initial goals. In working collaboratively to create metrics, you should ensure that they meet the SMART metric guidelines. This will help enable the systematic data-driven approach of the LSS methodology in the future. If you don't set the baseline, you will never know how well your improvement efforts are paying off; if they do not align with your goals, your leadership will never buy in.

Through the creation of EWPMs and with key stakeholders together in a room, you can easily identify your high-value initiatives (HVI). As part of your LSS deployment, you must ensure you are focusing on both quantitative and qualitative factors for project selection. EWPM enables you to do this easily by having a majority of the data at your finger tips. If you are at the beginning of a deployment you will most likely focus on lean initiatives that eliminate waste from your organization. This initial elimination of wastes is known as eliminating the ground fruit (see Figure 4.10). When you develop your initial set of project charters, please ensure that they are all directly correlated to the strategic plan you have established.

Types of initiatives	Types of methods used	Maturity of LSS program
Ground fruit – quick-win projects	Business case analysis, Kaizen	Infancy
Low-hanging fruit	Kaizen of LSS DMAIC	↓
Reduction in process variation or new process design	LSS DMAIC or DFSS	Mature

Figure 4.10 Correlation of LSS initiatives, methods, and maturity.

EWPMs also assist in supporting an environment of continuous process improvement or CPI. They accomplish this objective by enabling employees to openly provide input on a process. It is critical to the success of any CPI initiative to actively seek out process owner and stakeholder input (via EWPM meetings) and allow them to directly impact the processes that have an impact on their day-to-day work. Beyond fostering this environment, the CPI aspect comes into play by ensuring your "best practice" procedures are globally standardized and updateable in real time. Overall, leadership must foster an environment, throughout EWPM meetings, that rewards out-of-the-box thinking. This displaces the need for ad-hoc improvement efforts by motivating employees who are enabled to provide input directly to their supervisors and other leadership.

EWPMs can become your baseline for training requirements or even enhance your training capabilities. A significant percentage of the workforce is set to leave in the next 10-15 years due to the retirement of the baby boomers. This will be a great loss of intellectual capital and many organizations are concerned. EWPMs can help ensure that your organization is prepared for this loss by documenting knowledge before people leave the workforce. Your organization must take the time on the front end to prevent significant knowledge gaps. Providing repeatable, robust, and documented processes significantly reduces the impact of organizational turnover and ensures that the loss of a single person does not adversely impact your capabilities.

The easiest way to gain leadership buy-in is by means of measurable and sustainable results that directly map to leadership's key performance metrics. Nothing gains leadership buy-in better than showing results. EWPMs enable you to complete this seemingly monumental task by following a systematic approach to identifying low-hanging fruit that is directly tied to organization-wide goals. Most continuous process improvement projects fail due to a lack of leadership support; if you use EWPM in your approach, you can help guarantee success.

The level of detail for which you create your initial EWPMs has a direct correlation to the resources available. For example, you will not be able to map the policy processes if policy personnel do not attend your meeting. This is an example of why leadership support at the highest level is instrumental to your success (see Table 4.6 checklist).

Table 4.6 EWPM checklist.

- Leadership sponsorship
- Cross-functional SME teams
- Initial high-level maps
- Identification of HVIs
- Detailed maps
- Leaning
- Standardization
- Implementation

What is the preparation required for a successful EWPM meeting? The most important aspect of a successful session is ensuring that all necessary stakeholders are in the room. This does not simply mean having a representative from each part of the organization. The stakeholders at this meeting must represent every organization within your EWPM scope and must have an intimate knowledge of the process at the most detailed level. Otherwise, the output will not be at a level to allow you to use your EWPMs to identify initial process improvement initiatives.

When you have the necessary people in the room along with the necessary LSS facilitation supplies (markers, butcher block paper, and so on), you can begin creating your EWPM. The first thing you will do as a group is to map out the high-level enterprise view of the complete process from start to finish. At this point you will focus on identifying the high-level steps involved and the respective process owners. This information will be used in the next step to determine which stakeholders must be involved in developing the lower-level process maps.

To illustrate this, we will describe the process required for making brownies as an example throughout this exercise. This exercise provides a simplified understanding of the tools by means of a common process. Figure 4.11 illustrates a simple initial high-level process map.

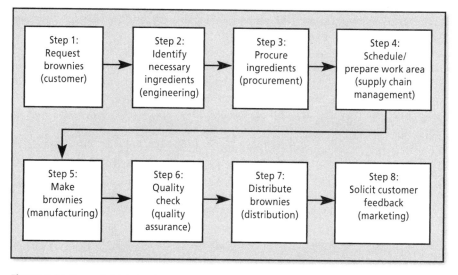

Figure 4.11 Example high-level process map.

As you can see from this simple example, many stakeholders are involved with even the simplest of processes. This is why you must have the right people in the room.

Once you have created the high-level EWPM, you can begin to create the detailed breakdown of each process step. It is good to have everyone in the room, even though they may not understand the sub-process being mapped; this is where you begin to identify opportunities for improvement. (For example, if procurement knew that customer service was receiving feedback on the poor quality of the frosting, they would be prompted to change the recipe or the process). This is where you also begin to lay the foundation for a future-state organization without functional barriers or stovepipes. We will begin to obtain our baseline for improvements by identifying quantitative measurements such as cycle and processing time. The metrics you obtain as part of your EWPM session should be directly linked to your strategic goals and objectives. Other measurements you might want to obtain as part of this effort include inventory levels, number of resources involved, queue time, and other metrics discussed in the balanced scorecard section.

Figure 4.12 is an example of the next level of the EWPM process. As you can see, we begin to see a task-level view of the process from an enterprise-wide perspective.

For the purposes of this example we will go no further into process detail. In a real-world environment, you should go to a task level of detail where you can begin to write a standard operating procedure (what steps are involved in preparing the line, what temperature is used to bake the brownies, and so on).

Beyond the process map shown in Figure 4.12, EWPM also provides an opportunity to visualize other flows. These flows include material, information, and transportation. They are commonly overlooked areas of improvement—material: inventory levels and work-in-process; information: manual or electronic; transportation: vehicle routing and load quantities. EWPM, when used in its entirety, forces you to look at your organization from a variety of angles, another reason why the tool is so valuable.

Keep these things in mind when you begin to map out your EWPM:

- Although a process may have a single process owner, it's likely that several stakeholders are involved. Take special care to ensure your EWPM is not constructed in a vacuum. By definition, you must ensure that you are involving a cross-functional team (suppliers, customers, and internal and external stakeholders across all functions) that will allow for the mapping and improvement of the entire value stream. Sometimes breakthrough improvements come simply from increased communication across functional boundaries. Also, take note of key stakeholders so you can accurately integrate them when creating your standard operating procedure (SOP).

- Ensuring you have the appropriate subject matter experts involved in your EWPM session is critical to the overall success of your standardization effort. If participants don't have an intimate understanding of the process, they will not recognize outliers to the normal process and it may be difficult to reap the full benefits of your EWPM effort. Also, if subject matter experts are not involved on the front end of the effort, they may not buy in to the implementation of improvements later.

Focus on Waste, Then Variation 91

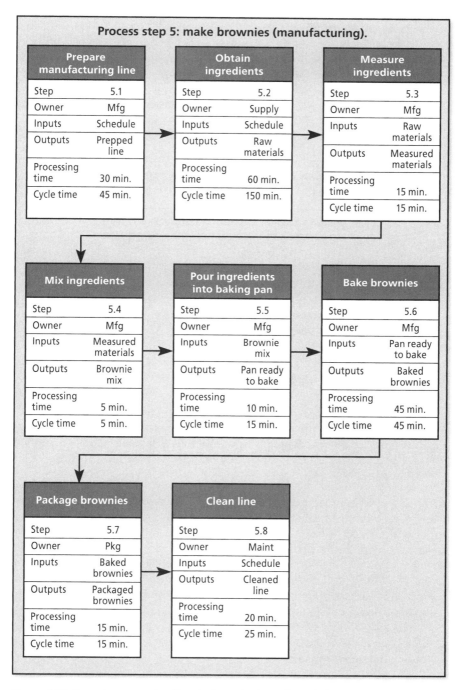

Figure 4.12 Example process step 5: make brownies (manufacturing).

- Prior to deploying standardized processes, you must create a sound sustainment plan. Consider who will be responsible for making updates, how updates will be requested, and how often they will be updated. Your deployment strategy should link back to the fundamental idea of continuous process improvement and allow for out-of-the-box thinking by process owners and process users.

- As you finalize your EWPMs, work to eliminate waste prior to standardization and implementation. Waste is any activity that requires time, space, material, information, equipment, and/or other resources but does not address true customer requirements or needs. (See Table 4.7.) Continually ask this question: "Does this process step add value to the customer?" If not, work to eliminate the step. Pairing your EWPM effort with a complete value analysis further enhances the systematic approach to waste elimination. EWPM focuses on the entire value chain, from the first supplier to the final consumer. I have found it beneficial to provide an end-to-end value stream example to present event attendees with an understanding of a value stream (from planting orange trees to drinking orange juice or from planting a cocoa bean to eating a candy bar). VSM is valuable because it allows the team, with all stakeholders involved, to see the waste such as rework loops, redundancy, excessive decision loops, and unnecessary hand-offs and obtain input from other functional organizations on non-value-added items.

- On-the-job training is necessary as part of any mapping effort to ensure the project team understands all of the steps involved (for example, SIPOC development, process mapping, value analysis).

Prior to implementing your standardized processes, be sure they are as lean as possible. Complete a value analysis prior to your deployment of the standardized process. If this is integrated into your initial EWPM meetings, with your process SMEs in attendance, this can be a fairly straightforward step. An overview of value analysis criteria is outlined in Table 4.8. These are normally integrated into an EWPM meeting after your initial current state map is constructed. Highlight each process step as value-added (VA), non-value-added-but-required (NVAR), or non-value-added (NVA); green, yellow, or red respectively. This is typically one of the most productive exercises in terms of eliminating large amounts of process waste.

Table 4.7 Types of waste.

Defects
- Manufacturing: A defect in the product produced.
- Services: An error or incompleteness of information.

Rework
- Manufacturing: Time spent trying to reclaim value from a defective product.
- Services: Not providing accurate requirements initially or incorrectness of data provided.

Over Production
- Manufacturing: Anytime more resources are engaged than absolutely necessary (for example, batch production).
- Services: Producing information before it is required or providing more information than necessary (such as unnecessary reports, extra copies).

Waiting
- Manufacturing: Queue time or any time spent waiting that does directly add value to the customer.
- Services: Processing information in batches (for example, waiting till quarter close to process all of "x" transactions).

Transportation
- Manufacturing: Relative to the finished product; any movement that does not directly add value to the customer (for example, movement between stations that does not add value).
- Services: Additional process steps that are not required or moving information from one place to another without adding value.

Intellect
- Manufacturing: Not soliciting feedback on the process directly from the process SMEs.
- Services: The underutilization of human capital (for example, not fostering creativity, high-morale, or continuous professional development of employees).

Motion
- Manufacturing: This varies from transportation, as it is focused on the movement of people and/or equipment.
- Services: Unnecessary approvals or sign-offs.

Excess Inventory
- Manufacturing/Services: Having excessive inventory relative to true customer demand (raw materials, work-in-process, and finished goods).

Table 4.8 Value analysis criteria.

> **Value Analysis**
>
> - Value Added – These are process steps in which value is directly related to a customer requirement. In the automobile industry, an example of this could be installing a drive shaft.
>
> - Non Value Added but Required – These are process steps that are required even though they do not directly add value to the customer. An example of this could be FDA compliance in the pharmaceutical industry.
>
> - Non Value Added – These are process steps that add no value to the customer. An example of this could be providing a hard copy of a monthly bank or credit card statement to customers who do not require them.

EWPM is a valuable LSS tool that can be used for a variety of things during a public sector deployment and sustainment—for setting a baseline and helping to identify the initial high-value initiatives, for empowering employees and increasing overall morale, and for significantly reducing waste and refocusing energy on value-added analysis resulting in increased leadership support. As you can see, much data has already been captured using this tool. As a secondary effect, this tool will put you on the path to compliance such as ISO certification and government policy compliance. All of these will positively impact your use of LSS as an organizational process improvement methodology and help you gain and sustain leadership support.

STANDARD OPERATING PROCEDURES

Impact in Overcoming Public Sector Challenges								
Breaking down stovepipes	Creating urgency	Leadership support	Metrics	Common goals	Increasing customer focus	Reducing impact of turnover	Overcoming complexity	Fostering collaboration
						X	X	

What value is there in standardized processes? We all think we know best how to complete the work we do. That may be so, but having documented standardized processes is an enabler for continuous improvement. Standardized processes force true LSS opportunities to become visible. The uncovering of process improvements through standardization must be seen as a positive by senior leadership. Otherwise, your organization will never want to document a truly broken process.

> **Why does process standardization matter?**
>
> Making widgets is easy, but Jim and Bob continue to get different outputs. Analyzing a budget can be easy, but Tom and Phil always end up with different numbers. These are just two examples of how a non-standardized process can cause significant organizational issues.
>
> Let's expand on the Tom and Phil budgetary example. Tom and Phil continuously come up with different numbers, which affects their organization's ability to accurately plan resources for the year. This occurs for various reasons. They may be using different data sets; one may be basing the budget on the accounting year, the other on the calendar year. One may be utilizing cost accounting, the other including overhead for maintenance and supplies. This is just a simple case of how variation in the process can affect the outcome of a value stream. A standardized process flow, with an accompanying standard operating procedure (SOP), significantly assists in ensuring a robust and repeatable process with standardized outputs. A non-standardized process can create significant amounts of waste in terms of re-work while also affecting overall organizational and team morale. Keep this in mind when institutionalizing standardized processes within your organization. Employees do not want to output a defective product or service, they simply do not have the appropriate tools or institutional knowledge to create the correct product or service.

Upon completion of a baseline SOP you can use the information to pursue perfection in your process, create a quality control system, and institutionalize government regulations, policies, and guidance by building them into the process.

SOPs help in reducing inaccuracies, poor quality, re-work, missing reporting requirements, and significant hours spent recreating processes. They provide a clear understanding of what must be done, when, by whom, and how; what the inputs are; what the output should be; how long it should take; and how often it should be done. They allow you to more effectively spend your time on analysis or fact-based decision making rather than on the transactional aspects of a process. Ideally, they should be centrally located, easily maintained, and globally updateable (true continuous process improvement).

The creation of SOPs is critical to implementing a standardized process and assisting in continuous process improvement and organizational continuity. Upon completion of your detailed EWPMs at the lowest levels, with involvement from the same stakeholders or SMEs, you can now begin creating your organizational SOPs. The fundamental purpose of an SOP is to ensure a repeatable and robust process that provides the same key process output time after time.

Table 4.9 illustrates a template that can be used to develop each SOP. Once you have developed your SOPs, post them at the location where the process is completed (in a manufacturing environment at the process step; in a services environment, on the process owner's/user's desk).

Table 4.9 Example SOP for process step 5: make brownies.

Overview of Process	The purpose of this process is to manufacture brownies in preparation for customer distribution.
Process Trigger	Scheduled manufacturing line
Frequency	Daily
Start	7:00 AM (EST)
Completion	3:00 PM (EST)
Process Owner	Manufacturing
Other Key Stakeholders	Supply, packaging, maintenance
Key Process Inputs	Oven temperature: 375° F Pan size: 10"x12" Pan shape: rectangular Pan color: clear Amount of batter per pan: 8 oz. Oven rack position: third position from top rack Pan position on rack: front-middle Cutting size of brownies: 1.5"x1.5" Temperature for packaging: < 80° F Packaging materials: cellophane
Process Steps	Refer to process map step 1.5 – making brownies
Impacts on Parallel Processes	Should only be completed when a request for process map step 1.1 is completed
Key Process Outputs	Packaged brownie

Here are few notes about creating organizational SOPs:

- An SOP is not needed for every process at the lowest level. Make an informed judgment about the number of SOPs required for your organization. For example, you may produce an intuitive report that takes you 15 minutes twice a year. You probably need not create an EWPM and SOP for this process. If you create a report that takes 2.5 hours every day and involves a large number of inputs, and stakeholders and significant process knowledge, it most likely requires an EWPM and SOP.

- The process to update your SOPs should be transparent and easily understood by everyone affected by it. If a change is approved by the process owner, you should have the flexibility to implement on a global scale in a short time frame. Leverage technology to accomplish your goals and communicate successes across the board (best practices). Documenting change is the first step in the improvement process. You should be moving your organization toward an environment of continuous process improvement in which employees are empowered to change processes by which they are affected.
- An organization *must* have an appropriate communication and change management plan to ensure all stakeholders understand the value of your EWPM and are aware of the capabilities it provides: global standards, feedback, and POCs.

SOPs take time to create, but they can help an organization in various ways from eliminating waste, defects, and rework to fundamentally improving morale through a feeling of process ownership and employee empowerment. You must ensure that these points are continuously stressed to and by leadership.

SPAGHETTI DIAGRAMS

Impact in Overcoming Public Sector Challenges								
Breaking down stovepipes	Creating urgency	Leadership support	Metrics	Common goals	Increasing customer focus	Reducing impact of turnover	Overcoming complexity	Fostering collaboration
							X	X

The purpose of a spaghetti diagram is to illustrate the movement of a process, product, or service. This tool can be used with equal value for manufacturing, maintenance, or transactional processes (such as moving from desk to desk). There are two key inputs required to create a spaghetti diagram, the steps involved as part of the process (such as a current state process map) and the physical layout (which can be built in a variety of software programs).

Figure 4.13 illustrates that even the simplest of processes can have excessive waste. The value gained in completing a spaghetti diagram is similar to that provided by most of the tools in the LSS toolkit. It allows project team members to visualize the true process. In this case, it will allow an organization to visualize transport waste, thus providing the opportunity to create a future state process that eliminates this waste.

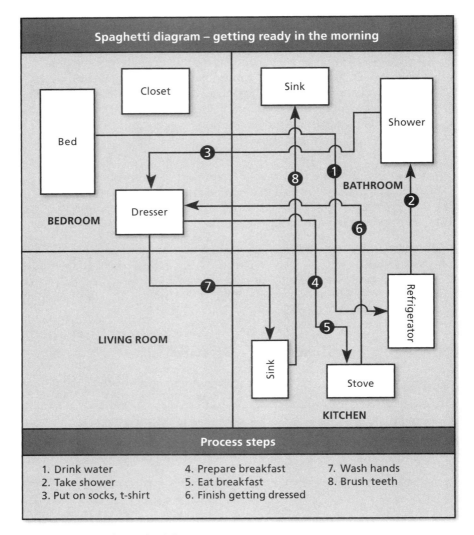

Figure 4.13 Example spaghetti diagram.

Figure 4.14 illustrates an example of how the waste of transportation can be removed from the earlier example. In improving the process, we were also able to combine "put on socks, t-shirt" and "finish getting dressed" as well as "take shower" and "wash hands."

Focus on Waste, Then Variation 99

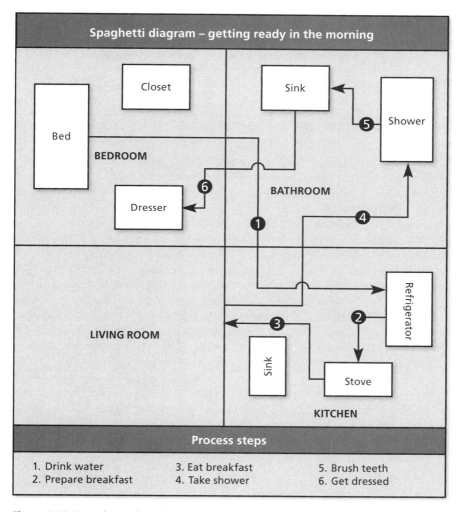

Figure 4.14 Example spaghetti diagram improved.

KAIZEN

Impact in Overcoming Public Sector Challenges								
Breaking down stovepipes	Creating urgency	Leadership support	Metrics	Common goals	Increasing customer focus	Reducing impact of turnover	Overcoming complexity	Fostering collaboration
X	X	X		X	X		X	X

Kaizen is a technique for continuously and incrementally improving across an organization. Think of the purpose as a mini-DMAIC event with the resulting improvements being implemented within 90 days of event completion. A typical kaizen event is accomplished in approximately five days; but more time than that is required for preparation.

Preparation for a solid kaizen event typically starts about 90 days in advance of the "in-residence" facilitation. The prep work begins with the identification and validation of an appropriate improvement event. This should be an outcome of the project selection, process mapping, or other project identification exercises completed previously. When an approved event has been selected, it's necessary to identify an appropriate champion. The champion must be willing to provide the resources (for example, people, money, data) and the excitement and support needed for implementing improvements in such a short time frame.

Once a project and a champion have been identified, the next step is to identify the appropriate kaizen event team. The target size for this team should be approximately 8-12 people; this will ensure that everyone has a voice while still allowing the facilitator to control the group. This team should encompass all functional skill sets as outlined as part of the project charter scope. In order to save time and focus the team, pay attention to a few things in advance of the event:

- Provide training on the tools that will be used (for example, process mapping, fishbone analysis, improvement plans).
- Secure sign-off and understanding of the project charter from all team members.
- Create a SIPOC with the entire team (via teleconference, if necessary) to identify and validate high-level suppliers, inputs, outputs, and customers and the start, end, and scope of the process.
- Develop a measurement plan for items identified as part of the project charter and SIPOC discussion for use throughout the kaizen event.

- Secure an offsite location to ensure no project team members are inadvertently called back to work or unable to focus 100% on the kaizen event.

With the pre-work completed you can begin to facilitate the actual event, which can vary slightly based on transactional vs. manufacturing activities or the type of improvements targeted. Figure 4.15 illustrates a sample agenda that focuses on mapping the current state for a transactional process, identifying possible waste and potential root causes, developing a future state process map, and developing an improvement/implementation plan. To ensure the entire group understands the process, use tools included in the Basic Quality Tools section of this book to identify and validate root causes (with data collected as part of kaizen pre-work), develop robust improvement and implementation plans, and ensure buy-in from all stakeholders involved.

	DAY 1	DAY 2	DAY 3	DAY 4	DAY 5
AM	Welcome kickoff • Expectations • Charter/SIPOC Current state mapping • Training • Walk the process • Collect process data	Root cause and value analysis • Training • VA/NVA/NVAR • Identify root causes	Mid-week vector check Improve • Training • Brainstorming • Improvement bursts • Affinity diagram • Multi-vote	Future state (cont.) Finalize improvement plan Implement • Create action plans	Implement (cont.) Next steps • Create action plan for incomplete steps • Create project plan • Develop meeting schedule
PM	Current state mapping (cont.) • Collect process data	Root cause and value analysis (cont.) • Training • VA/NVA/NVAR • Identify root causes	Future State • Training • Build future state • Improvement estimate data	Implement • Assign break-out teams • Draft materials "red-line" guidance policy	Outbrief

Figure 4.15 Sample kaizen agenda.

Keep in mind that you can mix and match the tools used in each phase of your kaizen event. For example, you may find it valuable to use current state mapping and spaghetti diagrams during the Measure phase. Think back to the DMAIC tollgate discussion at the beginning of this book; as long as you are defining, measuring, analyzing, improving, and controlling you are following a robust systematic process. The use of tools can cross multiple phases (for example, you can use brainstorming in almost any phase).

The key to the success of any kaizen event is the use of 30-, 60-, and 90-day action plans. In a manufacturing environment, the intent of a kaizen event is to complete improvements immediately during the event (move machinery, change procedures, and so on). In a transactional or public sector environment, this may not be feasible. Develop an action plan with clear ownership, accountability, improvement timelines (no actions should take more than 90 days), and 100% support from the project champion. This is necessary in order to complete any data validation, standard operating procedures, control plans, or policy/legal changes required.

Kaizen events can be instrumental in helping gain support at the beginning of any LSS deployment. They provide almost immediate results and this makes them highly visible. The events also help gain commitment and buy-in from team members who support the success or failure of the LSS program overall. The goal during each event should be to gain savings or identify improvements and also to provide momentum or create a positive "buzz" about LSS methods.

Public Sector Project Highlight
Kaizen Events, More Than Just Improvements
During a recent public sector kaizen event, significant improvements were made relative to reducing waste, defects, and cycle time. These were not the only benefits of the event. It also increased communication between organizations and identified further opportunities for improvement. Kaizen events provide value in terms of the improvements gained; they also help break down organizational stovepipes, foster an environment of increased collaboration, and help gain sustained buy-in and support through "ah-ha" moments.

5S

Impact in Overcoming Public Sector Challenges									
Breaking down stovepipes	Creating urgency	Leadership support	Metrics	Common goals	Increasing customer focus	Reducing impact of turnover	Overcoming complexity	Fostering collaboration
X	X	X		X		X	X	X

Have you ever wasted time looking for your car keys or cell phone? Similar things can occur in both public sector service and manufacturing environments. Time and energy are wasted every day searching for a torque wrench, stapler, office supplies, or electronic file in the "black hole" that may describe your hard drive. 5S is a tool that can help eliminate this waste and ensure sustainment of a 5S organization in the future. See Table 4.10.

Table 4.10 Fundamentals of 5S.

Sort	Eliminate any items that are deemed unnecessary to complete a process.
Set in order	Organize all items that are necessary to complete a process.
Shine	Ensure necessary items are always clean and ready for use in the completion of a process.
Standardize	Standardize and leverage best practices by communicating the first three "S's" across the enterprise.
Sustain	Create a sound sustainment/control plan to ensure the continued success of the first four "S's."
Safety (the 6th S)	Ensuring safety of personnel and customers should be fundamentally integrated within each of the areas above.

Sort

In the sort phase of the 5S process attention is paid to eliminating items that are not necessary in the focus area (outdated electronic files, unneeded tools, supplies that are used infrequently, and any other item that is not required to complete the process). The value is two-fold: 5S eliminates items that clutter the work area and items that could ultimately result in a defect (using the broken stapler causes rework and wastes resources). Items that are deemed unnecessary should be red-tagged and moved to a location off to the side. If the focus area is an electronic file structure, then unnecessary files should be moved to a folder for possible deletion.

Once items are red-tagged or moved to a folder for deletion, a 30-day grace period begins. If tagging errors are not discovered during this time, the items should be auctioned to other areas of the organization. *Auctioning* simply means allowing another person or department to take ownership of the items or files. Someone in authority in the new area must agree that the part or file is necessary for the completion of a process. Otherwise, you're simply moving waste from one area of the organization to another. Items that are not transferred via the auction process should be sold, recycled, given away, or deleted/discarded. This helps eliminate the "noise" that results from clutter or unnecessary items.

A typical pitfall to the sort phase is that people believe they may need something again in the future. These items should be red-tagged and moved to the auction area. If a person is unable to provide evidence as to how a particular item is used consistently as part of the process, it should be eliminated from the process.

Set in Order

Have you heard that old expression? "A place for everything and everything in its place." In simple terms, *set in order* means tools, supplies, and equipment have a designated space, whether in the office or on the shop floor, and that everything is kept in its place or returned to its place after use. Now that you have eliminated unnecessary items or files, be sure everything that remains has a specific use or storage location. This technique helps eliminate defects caused by utilizing the incorrect item (think of version control on an electronic file or using a regular mallet rather than a rubber mallet) and increases the speed with which necessary items can be located (helping to reduce cycle times). This visual organizational strategy will also help you recognize immediately when something is missing.

A hospital operating room is an extreme example of the value of setting things in order. It is difficult to lose a scalpel or other piece of medical equipment in a patient if an outline of the item on the operating table provides a visual indication that it is missing.

Shine

Now that you have eliminated unnecessary items and ensured that everything remaining has a permanent location, you must ensure everything is clean and ready for use at all times (tools lubricated and calibrated as needed, the stapler refilled, and the hard drive purged regularly to maintain your files). This is a fairly intuitive part of a 5S initiative; the most critical aspect in this step actually coincides with the next two areas, standardization and sustainment of the first three improvements.

Standardize

You have the correct materials for the job, you can find them quickly, and they are ready for use. The next step is to standardize best practices with relation to the first three "Ss" across the rest of your organization. This can be accomplished through standard operating procedures and the clear communication of expectations to other business units. Once you have standardized the outcome of sorting, organizing, and shining in the first three steps, the final step is sustainment.

Sustain

There is nothing worse than improving a process and then seeing it go back to previous unsatisfactory performance levels. 5S is no different. You must create a sustainment plan, with identified points of contact (typically the process owner in 5S), to ensure that your initial successes become lasting improvements.

Safety

In traditional lean methods there are only five "Ss." Safety is assumed. Because personnel and customer safety is essential, it is beneficial to have it as a sixth "S." This ensures that your improvement team explicitly looks at all improvements with safety in mind. Implementation of change must not negatively affect safety (although you might not use the fire extinguisher often, it is essential on your factory floor). In fact, safety must be increased whenever possible.

5S Events – How do I implement 5S?

Similar to kaizen events, 5S events encourage rapid change by involving all stakeholders on the front end. To facilitate a successful 5S event, follow this step-by-step approach:

1. Identify the focus of the 5S event. This could be a manufacturing line, a supply cabinet, departmental work desks, an organization's electronic shared drive or team room, or any other aspect of the organization that would benefit from being less cluttered, more organized, cleaner, standardized, or safer.

2. Identify key stakeholders who should be involved in the event. Gain sign-off from the project champion to ensure that resources are available and focused (meaning no calling out personnel for "more important issues" during the 5S event).

3. Obtain necessary 5S event facilitation materials. These materials typically include red tags, yellow tape, label gun, cleaning supplies, calibration tools, standard operating procedure templates, and sustainment plan templates.

4. Schedule the event and send an agenda to all attendees along with a brief description of the event expectations and the 5S tools (see the sample agenda and timeline illustrated in Table 4.11). It is typically easier to complete the 5S before the start of a normal work day (before starting production).

Table 4.11 Sample 5S agenda and timeline.

Overview and expectations	8:00 – 8:30 a.m.
Sort	8:30 – 11:00 a.m.
Set in order	11:00 a.m. – 12:00 p.m.
Shine	1:00 – 2:00 p.m.
Standardize	2:00 – 3:00 p.m.
Sustain	3:00 – 4:00 p.m.
Safety (the 6th S)	4:00 – 5:00 p.m.

The sample agenda is subject to the size and scope of your initiative, but it provides an example of how quickly 5S can be implemented in an organization. Please note that auctioning of your red-tagged items should not slow progress. You can immediately move into the other 5S phases and hold your auction 30 days from the day of your 5S event. Note that only one hour of the day is formally dedicated to safety. Please let everyone know that safety should be integrated within each improvement step.

5. When you have set the agenda and gathered attendees in the area that will be the focus of the 5S event, set expectations for everyone. If possible, ask your project champion to assist in setting the expectations.

6. Move directly into sorting your organizational assets specific to the scope of the 5S event. This is where you will typically spend the majority of your time. People fear change and find it difficult to identify wasteful items and red-tag them for elimination.

7. Next, identify a permanent location for each item and mark it as such. One way to accomplish this is to mark the area around each item with yellow tape and create a label for it.

8. Then be sure everything is clean and ready for use. Be sure tools are clean and calibrated; be sure office products such as staplers and printers have adequate supplies.

9. After the first three "S's" are improved as much as possible, create a standard operating procedure and communicate across your enterprise.

10. Aside from safety, which we have already addressed, sustainment of improvements is the final step in a 5S event. You can accomplish this by completing a sustainment plan template as outlined in the kaizen event chapter and identifying a specific point of contact (normally the process owner) to be responsible for the sustainment of the results.

11. Finally, do a sanity check with your team. Review all actions to ensure that none of the improvements affects the safety of your organization's personnel, suppliers, or customers. Ensure that each item within your 5S scope is as safe as possible, reducing the risk of injury.

Reward team members who complete a successful 5S event; communicate results across the entire enterprise and highlight the speed with which you improved your work area. This will help increase the "buzz" relative to the value of LSS.

FUTURE STATE PROCESS MAPPING

Impact in Overcoming Public Sector Challenges								
Breaking down stovepipes	Creating urgency	Leadership support	Metrics	Common goals	Increasing customer focus	Reducing impact of turnover	Overcoming complexity	Fostering collaboration
X	X	X	X	X	X	X	X	X

Once you have developed the current state process map, facilitated a value stream mapping session, and identified potential improvements, it is time to begin mapping the proposed future state process. The first step in this process is to eliminate all non-value-added steps as identified in the value analysis. We will use the previously discussed current state process as an example. (See Figure 4.16.) As you can see, the measurement of the ingredients has been identified as a non-value-added step based on the Vo"X" you acquired earlier. (In the facilitation of your value analysis use green for value added, yellow for non-value-added but required, and red for non-value-added.) The supplier for these ingredients actually has an option available where the ingredients are pre-measured. This also demonstrates an example of how involving all your internal and external stakeholders as part of an improvement initiative can help gain significant returns.

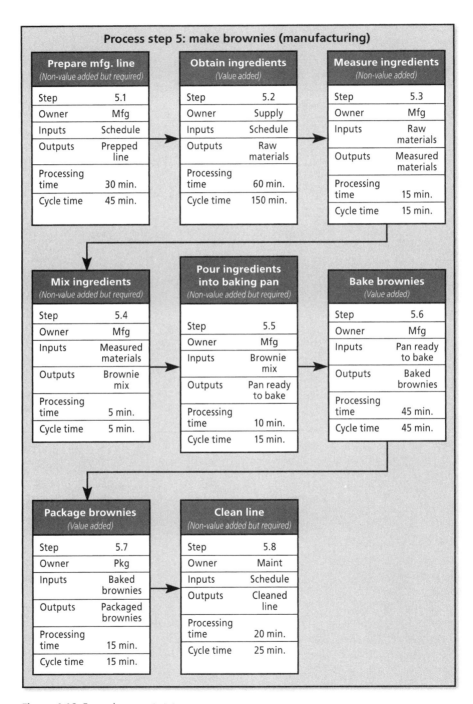

Figure 4.16 Example current state process map.

The next step in the future state mapping process is to investigate each non-value-added step in detail with your project team. Focus on "why must certain steps remain?" Is it a compliance issue, a legal issue, a safety concern, or another issue that requires the process step? Although a new process may sometimes result in some risk, you never want to risk the safety of your people. Remind your team that they must clearly state why non-value-added but required steps remain as they are outbriefing senior leadership.

In Figure 4.17, the project team and your suppliers identify the ability to eliminate the "mix ingredients" process step. The core supplier informs you of a new product being developed that provides customers with premixed ingredients.

Keep in mind the three major aspects that impact every organization: systems, processes, and people. (See Figure 4.18.) As you brainstorm improvements, take special care to ensure you are working to improve all three areas. As you consider each of these improvement areas, pay attention to potential return on investment. If it will cost $5 million dollars to implement a new information technology solution to eliminate a process step that takes five minutes, return might not offset the investment required.

In our scenario, the supplier has also informed the team that it is developing a "ready-bake" option that will combine premixed ingredients, based on customer specifications, with a disposable baking pan. At this point in time we should only eliminate the "mix ingredients" process step, but we should clearly articulate in the leadership outbrief the potential to eliminate another process step in the future. The "prepare manufacturing line" and "clean line" process steps cannot be eliminated because they address the quality control aspects regarding a robust and health code-compliant manufacturing line. However, we were able to eliminate two steps from the process, giving us the streamlined future state process illustrated in Figure 4.19.

The final step in future state process mapping is one of the most exciting, the calculation of savings. In this example we achieved the following results:

- Elimination of two process steps, which results in an easier to understand process and the ability to further minimize the opportunity for defects (the fewer steps there are in a process, the lower the likelihood of defective products).
- Elimination of 20 minutes of cycle time and processing time. If this process is repeated multiple times throughout the day, returns could be significant. (For a process completed 10 times per day, 5 days a week, and 50 weeks out of the year, the time saved could be as much as 833.33 hours.)
- As a secondary effect, you have also most likely built stronger relationships with your suppliers and helped foster an environment of continuous process improvement within your organization.

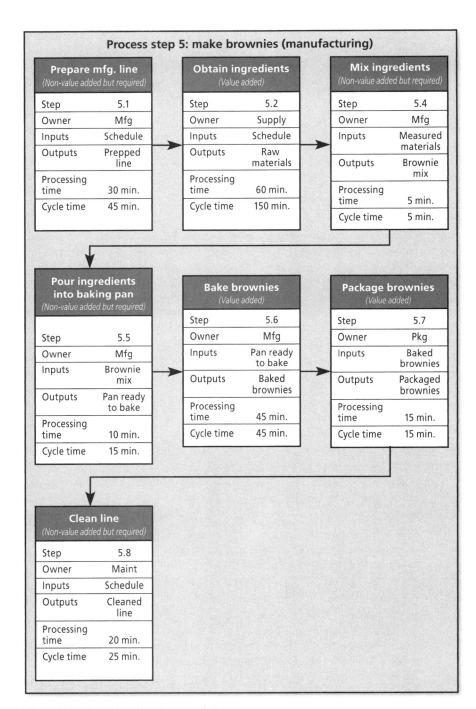

Figure 4.17 Example value stream analysis.

Focus on Waste, Then Variation 111

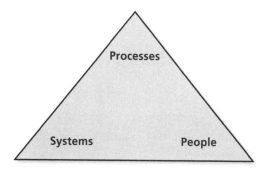

Figure 4.18 Three major aspects that impact every organization.

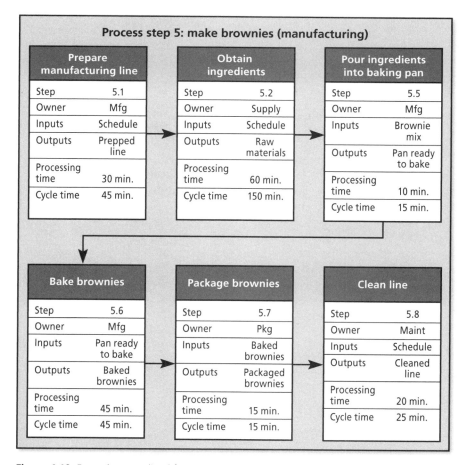

Figure 4.19 Example streamlined future state process.

It is important to recognize the team as you are briefing the results to senior leadership. Congratulate the team publicly; the results gained would not have been possible without their hard work, dedication, and subject matter expertise. This is an opportunity to build and sustain momentum with the executive steering committee, project sponsors, and champions and communicate success across the entire enterprise.

> **Public Sector Project Highlight**
>
> **Future State Pilots**
>
> As described earlier, LSS programs often encounter a fear of change. Because LSS is fairly new to the public sector and there are residual fears due to unsuccessful improvement and efficiency programs, this fear is prevalent. One approach to overcoming this fear is by proving results in a safe environment. This can be accomplished easily by piloting proposed improvement solutions. Pilot projects not only help reduce cost and risk, but also increase the likelihood of success and assist in overcoming fear of change.

LEVERAGE FAST, VISIBLE, HIGHLY IMPACTFUL SUCCESSES

Upon completion of the first future state process mapping session, this is typically an excellent time to implement some fast, visible, and high-impact quick wins. The goal is to display the value of the LSS toolkit and its potential to significantly affect the organization with input gained from a small number of tools (project charter, SIPOC, current state process map, value analysis, future state process map, and an improvement/implementation/sustainment plan). These initial successes will not only secure buy-in from senior leadership, but will also create a "buzz" across the organization by gaining recognition for team members and other personnel involved. This multiplier affect will encourage future stakeholder involvement.

> **Public Sector Project Highlight**
>
> **Standardize the Savings Process**
>
> There are cost savings and then there is cost avoidance. Although people typically focus on cost savings, cost avoidance is just as crucial in the public sector (based on the fiscally constrained environment). The goal is to create a standardized process to quantify both types of cost reductions (remember savings in the public sector can be in the millions of dollars). I have seen organizations accomplish this with a simple spreadsheet. The key, of course, is to use a template that is intuitive and robust (that is, it can be used across the enterprise), and training and communication plans that are deployed to all stakeholders (integrated as part of awareness training). It is critical that the linkage to the funding re-allocation plan is also clearly communicated (that is, we are not taking away money or resources). When a robust, standardized approach to quantify results is used, light bulbs go on when results are gained. Because the results are quantified, leadership will be more apt to provide additional resources to the LSS program. When leadership takes a positive position, it creates excitement across the enterprise (people tend to rally around their bosses).

SPOTLIGHT – FACILITATING EFFECTIVE MEETINGS AND EVENTS

The facilitation of effective meetings and events is instrumental to the success of any LSS program. This is the opportunity to demonstrate how effective and valuable the various tools can be, while also beginning to gain buy-in for both the overall program and the specific project or initiative. Keep in mind several things when facilitating meetings or events:

- Always set goals for the meeting. If a targeted outcome is not defined and communicated to all stakeholders in advance of the meeting, it may be difficult to keep the meeting on track.
- Research the audience. If they typically hold meetings at 7 a.m., work with them to meet this timeline. If they typically allow time for a few short breaks rather than a one-hour lunch break, follow that timeline as well. Although leadership support has been provided to facilitate the LSS projects and meetings, it is the responsibility of the meeting facilitator to engage and maintain an audience.

- Based on the goal for the meeting, distribute a detailed agenda and other necessary material in advance. This will allow all parties to come prepared. There is no value in surprising attendees with the final agenda, and this will likely result in excessive questions or a lack of support.
- During the meeting, follow the agenda from both a topic and timing perspective. If there is a need to run over or change topics from the original agenda, proactively gain buy-in from the group to make those changes. If areas of the agenda can be eliminated, communicate this to the group as well. There is nothing worse than sitting through a four-hour meeting simply because it was set to be a four-hour meeting, even though all the goals were achieved in the first two hours. Respecting the agenda will make stakeholders feel as though their time is valued.
- Upon completion of the meeting, validate that goals were achieved or an appropriate action plan created. Distribute action plans to all attendees along with minutes from the meeting. This will allow stakeholders to provide feedback on topics that may have been misinterpreted or allow for follow-on questions. If additional meetings become necessary at a later date, gain buy-in for the appropriate method for communicating this as well as tentative dates.
- As action items are closed out by the assigned meeting attendees, communicate news of this to the entire group. This will help to keep stakeholders both informed and engaged.

If the above items are considered as part of every meeting, there will be a lot more successful outcomes. It is the responsibility of the facilitator to ensure these items are integrated in each and every meeting or event.

Section 5
Basic Quality Tools

"If you can't explain it simply, you don't understand it well enough."

– ALBERT EINSTEIN

There are many tools in the LSS toolkit. The goal is to use only the tools necessary to reach a validated, fact-based decision. No bonus points are provided for using all of the tools, or the more advanced tools, when the simplest ones will work. This is not to discount the value and power of some of the more advanced statistical tools such as designed experiments, multi-vari charts, and regression analysis. The focus of this section will be to provide an introduction to some of the tools available, all of which have weathered the test of time and have proven they can yield amazing and lasting results. From my experience, it will be more effective to gain leadership support if leaders can easily understand the LSS process, the LSS tools, and the project outputs.

Some senior leaders or process owners may feel uncomfortable leaving the analysis to their process improvement teams. Some feel as though they already know the solution and that the analysis is simply wasting time and energy.

Let's look at a simple example. You are brushing your teeth one day when suddenly you feel a sharp pain in your head. What should you do? One response to avoiding the pain might be to stop brushing your teeth... forever. No one would want to be within 10 feet of your breath, and this would probably not actually eliminate the root-cause of the problem. Additionally, implementing decisions that have no data to support them can cause additional problems (plaque build-up, gingivitis, tooth loss) and you are likely to incur additional costs (for dentures) while still not identifying the root cause.

So what should you have done? This is where the LSS tools come into play. You could have collected data on when and where the pain occurred, the severity of the pain, and so on, and then used the basic quality tools to identify the root cause. If you had used Pareto analysis, the 5 Whys, and fishbone diagram tools you would have identified that the root cause of the problem was a small burn inside your mouth that was irritated by the toothbrush. You would have purchased a simple product at the grocery store for $5 and avoided a visit to the dentist, losing all your teeth, and unnecessary additional costs. This simple example shows the value in analyzing your data prior to moving forward with the implementation of improvements.

MEASUREMENT PLAN

Impact in Overcoming Public Sector Challenges								
Breaking down stovepipes	Creating urgency	Leadership support	Metrics	Common goals	Increasing customer focus	Reducing impact of turnover	Overcoming complexity	Fostering collaboration
X	X		X	X			X	X

Before we can discuss the basic quality tools, we must first ensure the appropriate data is being collected. The easiest way to ensure accurate, timely, and relevant data is to use a robust measurement plan. Keep in mind that all data should be linked to a key performance input/output indicator (KPI/KPO), CTQ metric, or potential improvement area. (See Figure 5.1.) This helps validate the linkage between the goals for a specific project to a specific Vo"X" item.

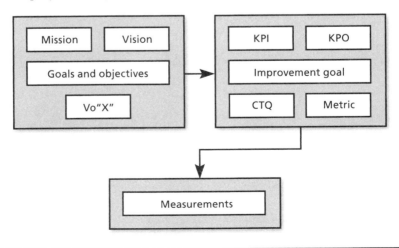

Figure 5.1 Measurement plan linkages.

A solid measurement plan typically includes the following data, although it can be tailored to a specific organization as appropriate:

Type of data – measurement type, discrete or continuous (see Table 5.1).

- Continuous or non-attribute data is collected on a continuum. One example of continuous data is time, which has an infinite number of possibilities (10:01, 10:02, 10:03 or 10:00, 10:10, 10:20).
- Discrete or attribute data has specific defined outcomes. "Yes or no" or multiple choice (A, B, C, D, E) responses are examples of discrete or attribute data.
- The type of data used should be dependent on the type of project. Continuous data can provide more detailed feedback on a specific process, but may take longer or be more difficult to obtain based on the maturity of an organization's measurement system. The key at this point is deciding what type of input is required for the project and ensuring that adequate time and resources are provided to complete the data collection effort.

Table 5.1 Example discrete and continuous measurements.

• Pass/fail (discrete)	• Lead time (continuous)
• Yes/no (discrete)	• Queue time (continuous)
• Time/date (continuous)	• Processing time (continuous)
• Cycle time (continuous)	• Value-added time (continuous)
• Inventory levels (continuous)	• Takt time (continuous)
• Machine up/down time (continuous)	• Defects/defectives (continuous)
• Customer service levels (discrete)	

Relationship to the project – Is the data specific to a goal, objective or improvement area/target, CTQ, or KPI? Always keep $F(x) = y$ in mind when developing your measurement plan. That is, all outputs are a function of the inputs. If you have variation in one input, it could create a multiplier effect causing significant defects or issues with the final output. In the example below, the call answer speed has the most impact on call abandon rate and overall customer satisfaction. (See Figure 5.2.) Therefore, measurements related to call answer speed should have priority.

	Outputs (Y's)	
Inputs (X's)	Call abandon rate	Customer satisfaction
Answer speed	●	●
Years of experience		○
Training		○
Correlation		
Strong	●	
Medium	○	
Weak	△	

Figure 5.2 Call answer speed.

Target sample size – It is not always cost or time efficient to use a population of data (all data). The outcome of this approach is inferential statistics (that is, we are inferring an outcome based on a limited sample of the entire population). As a baseline, a sample size of 20-30 typically meets most requirements, but in some instances this may still be too large (for example, $ spent per year). In cases where there is not enough data to reach a sample size of at least 20, it may be necessary to use historical data. Although acquisition costs may be higher, a larger sample size can provide increased confidence. If you have trouble identifying a robust sample, please refer to an appropriate text on calculating sample sizes to determine a minimum sample size for your specific needs. It is also beneficial to ensure your sample is random meaning that it represents a mix of all potential outcomes. This can be accomplished through the segmentation process as discussed earlier.

Collection method – Will the data be collected from an existing system, using surveys or another data collection method?

Output format and data requirements (for example, column A fiscal year, column B "yes/no" response, MS Excel, MS Word, and so on) – This helps ensure that the data can be quickly analyzed and used for the basic quality tools. Failing to accurately define the format for data output on the front end can lead to significant rework (for example, migrating MS Word data to MS Excel) and may even result in having to re-collect the data.

Translation requirements – It is important to make the entire team aware of what the output calculations may include (cycle time, pass/fail, and so on).

Due date – When is the data required?

Point of contact – Who is responsible for obtaining the data?

Now that we have acquired the necessary details for a robust measurement plan, an easy-to-follow matrix is provided in Table 5.2 to support future leadership approval requirements.

Table 5.2 Example measurement plan.

Measurement plan for Project XYZ							
Measurement	Data type	Sample size	Relationship	Format	Translation	Due	POC
Met customer requirements	Discrete	30	Customer satisfaction	"Pass/fail," customer name, date (mo/day/yr)	Customer satisfaction rating	1-Oct	John D.
Calls per day	Continuous	30	Capacity requirements	Number of calls by employee, date (mo/day/yr)	Cycle time	15-Oct	Jane D.

As you progress in the LSS methods, it is important to understand the accuracy of your measurement system (for example, are the measurements the same from one data collector to the next; is the gage calibrated correctly). This is completed by using tools and techniques such as measurement system analysis (MSA) or gage R&R (repeatability and reproducibility). At this point, we will focus on ensuring robust data is collected. As an organization becomes more mature in the LSS toolkit, it may provide value to expand into these MSA or gage R&R tools as part of every measurement plan.

> **Public Sector Project Highlight**
>
> **Measurement Plan Standardization**
>
> As part of a recent LSS initiative, a Master Black Belt noticed that one area was achieving a near best-in-class cycle time. Because this is such an outstanding achievement, the MBB decided to investigate the process in further detail to determine how they were achieving such great efficiencies. Upon documenting the process, the MBB realized that they were not measuring cycle time the same way other organizations measure it and they were not including the final review and decision loop. Although they were showing amazing efficiency gains, they were not really comparing "apples to apples." This is just one simple example of why it's important to ensure standardization in the measurement process. If everyone is not collecting the data and calculating the outputs the same way, it will be difficult to define which organizations need improvement and which are achieving excellent results. This can be overcome by ensuring the measurement plan includes communication with regard to how the measurements will be collected, when they will be collected, and why the data is being collected. The "why" is the most important piece supporting change management.

BENCHMARKING

Impact in Overcoming Public Sector Challenges									
Breaking down stovepipes	Creating urgency	Leadership support	Metrics	Common goals	Increasing customer focus	Reducing impact of turnover	Overcoming complexity	Fostering collaboration
	X		X	X	X			

Have you ever wondered how you are performing relative to other organizations? If so, benchmarking is the right tool for the job. Benchmarking current state performance and comparing it to better performing organizations allows you to identify a specific target. A benchmark can be identified one of two ways:

- Internal to your current organization (other divisions offering similar products or services), or
- External to your organization (direct competitors; other government contracting, logistics, manpower agencies; industry trends both in the private and public sectors; or best-in-class organizations across all industries and sectors).

It is oftentimes beneficial to benchmark organizations outside of the public sector to identify truly radical process improvement opportunities.

Benchmarking is beneficial in that it provides an achievable goal (another organization has already proven the ability to reach a given performance level). If possible, you should try to benchmark organizations that are willing to openly provide their performance data, lessons learned, and ideas for improvement (think of this as free advice to achieve improved performance). Obtaining this information can be very difficult in the private sector, but when you are a public sector entity, both private and public sector organizations are more apt to cooperate. Similar to Vo"X," this information can be obtained by interviewing and surveying industry experts at benchmarked organizations and through trade publications, websites, seminars, and site visits (site visits provide access to both personnel and the operating environment). The key is to properly scope the process, product, or service you plan to benchmark, identify an appropriate organization, and attempt to create an open communication channel with that organization.

Public Sector Project Highlight

Benchmarking

In completing a recent process improvement initiative, the team realized that the process required streamlining and that the product (a congressional reporting requirement) contained significant waste (as identified through a Vo"X" exercise). The team decided to benchmark other organizations' report submissions including the format for budget documentation across all public and private sector industries. As demonstrated in this example, benchmarking is not only for performance measures. It also can be used for reporting requirements, product specifications, or any other area where a clear comparison can be made to a benchmark product, service, or process.

RUN CHARTS

Impact in Overcoming Public Sector Challenges								
Breaking down stovepipes	Creating urgency	Leadership support	Metrics	Common goals	Increasing customer focus	Reducing impact of turnover	Overcoming complexity	Fostering collaboration
X	X			X			X	

Run charts are a great tool for providing trend data to identify data patterns. The information collected as part of a run chart can be used to see whether a certain person, operator, shift, location, or other performance measure is causing significant variation. If possible, we will shoot for approximately 20+ data points, given data availability. This should be adequate information to calculate an initial average and trend. (See Figure 5.3.)

1. The first step in creating a run chart is to collect the data. Similar to the previous discussion, we want to ensure this is a random sampling of data to help provide a robust inference of the total population (inferential statistics).

2. The next step is to plot the data gathered on an X- and Y-axis chart. The X-axis in this instance will be continuous data (run charts are not a good tool to use for attribute data) and the Y-axis will be the individual measurements.

3. Now that we have the data plotted on the chart, we will calculate and draw an average line (simply add up all the data points and divide by the sample size).

4. Once all the information has been plotted and the average line calculated, the final step is to interpret and maintain the chart. We will identify significant variation (use the information included within the Control Chart section to help determine significant variation) in the data and continue to add new samples to the run chart.

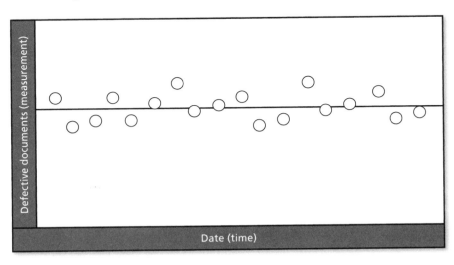

Figure 5.3 Example run chart.

If this were the run chart for an actual process, the variation would be considered fairly limited. There are no significant trends visible in the data; variation is not a concern as long as customer specifications are met consistently. This is not to say that waste or defects may not be a concern (based on performance metrics as defined in the balanced scorecard, 100 defective documents per day might be either adequate or poor performance). This is why it is critical to define thresholds for operational performance and an appropriate operational definition.

FISHBONE DIAGRAMS

Impact in Overcoming Public Sector Challenges								
Breaking down stovepipes	Creating urgency	Leadership support	Metrics	Common goals	Increasing customer focus	Reducing impact of turnover	Overcoming complexity	Fostering collaboration
	X						X	X

Did you ever have a younger sibling, niece, or nephew who continuously asked "why" to everything you said? At the time, you might have found this frustrating or annoying, but asking "why" is the fundamental premise behind the 5 Whys technique. With this tool, a user continues to ask "why" until the true root cause of a problem is reached. This simple approach can then be used to create a fully developed fishbone diagram.

The first step in the process is to identify the symptoms you are trying to drive to a cause (involving your key stakeholders, of course). Symptoms may be identified in the output of Pareto charts, correlations identified in scatter diagrams, non-normal distributions identified in histogram analysis, and any other areas in which you feel a root cause may exist.

In order to understand how to use this tool, let's use a simple example: your laptop computer won't turn on. (See Figure 5.4.)

If you had stopped earlier in the process, at a symptom rather than the root cause, you might have replaced the computer battery. This would have cost you money without actually fixing the problem. As you can see in this example, we only asked the question "why" four times. There is no magic in the number five. The critical concept is to continue asking "why" until you reach a "true" root cause. In this example, replacing the fuse corrected all the other symptoms. Continue to use the 5 Whys approach until you have explored all possible opportunities for improvement.

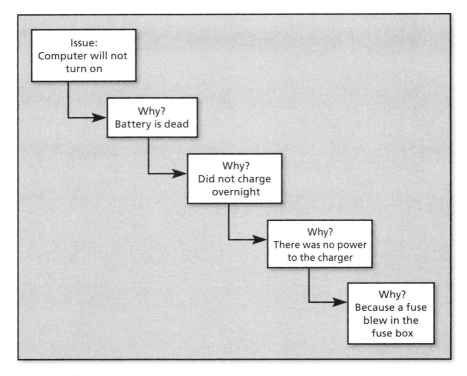

Figure 5.4 Five whys example.

Once you have exhausted your list of potential areas for improvement, begin to plot your 5 Whys output onto a fishbone diagram. Some practitioners like to group potential root causes by defect error (man, machine, and so on), but I find it more beneficial to categorize them by primary defect areas. This means putting the highest defect areas or largest potential opportunity towards the "head of the fish." Figure 5.5 illustrates how this simple tool can be used to identify true root causes. The root cause in this example is the overloaded fuse box, as it is the final cause in this cause-and-effect node.

The major caveat with this tool, as with many others in the LSS toolkit, is that you must have data to validate potential improvements or defect areas. (One way to validate a fishbone diagram is to use scatter diagrams, Pareto charts, or histograms). This will help ensure that you are solving problems, rather than symptoms.

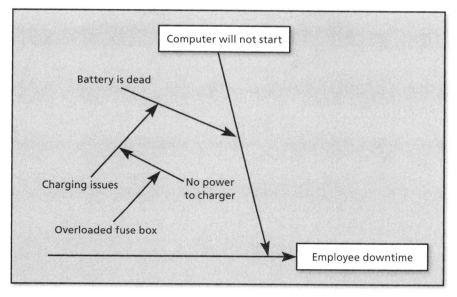

Figure 5.5 Example fishbone diagram.

Public Sector Project Highlight

Excitement for the 5 Whys

As part of a recent public sector kaizen event, the LSS MBB had difficulty engaging the team until they began to focus on the 5 Whys approach. Teams tend to become more engaged when they can provide active feedback quickly. There is nothing more fast-paced than trying to break down an opportunity for improvement with input from 8 to 10 people using the 5 Whys approach. This technique works to actively engage the team and gives everyone a chance to communicate potential opportunities and causes, making implementation that much easier for everyone involved. By a wide margin, change management was listed as one of the key success factors uncovered in a series of public sector interviews. It is important to leverage every opportunity to engage stakeholders and make them feel as though they have a real voice in the LSS process.

CHECK SHEETS

Impact in Overcoming Public Sector Challenges								
Breaking down stovepipes	Creating urgency	Leadership support	Metrics	Common goals	Increasing customer focus	Reducing impact of turnover	Overcoming complexity	Fostering collaboration
			X	X			X	

A check sheet is probably one of the easiest tools to understand in the LSS toolkit, involving simple hash marks to record the occurrence of discrete data. This tool is best applied when there is a small amount of simple data to be collected. The output of this measurement technique can be used as part of Pareto charts, histograms, scatter diagrams, and other basic quality tools. Figure 5.6 illustrates a variety of unrelated examples that assist in visualizing the check sheet approach to data gathering:

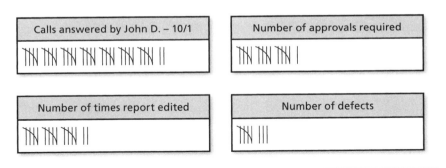

Figure 5.6 Check sheet approach to data gathering.

CONTROL CHARTS

Impact in Overcoming Public Sector Challenges								
Breaking down stovepipes	Creating urgency	Leadership support	Metrics	Common goals	Increasing customer focus	Reducing impact of turnover	Overcoming complexity	Fostering collaboration
	X		X			X	X	

Now that the process has been improved, there is no need to continue data collection efforts, right? Wrong. In fact, improving a process is only the first step to sustaining organizational excellence. Control charts are a powerful tool for continuing to improve a process and should be integrated whenever possible into your organizational dashboard.

To begin you must collect data, identify X and Y axes, and plot the information on a chart. Then you must calculate control limits and determine which control chart to use (see Figure 5.7 for attribute data control chart options). Control limits should not be confused with specification limits that are determined by the customer through the Vo"X" process (a specification limit is dictated by the customer, whereas a control limit is statistically calculated).

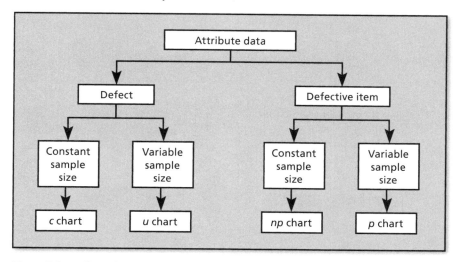

Figure 5.7 Attribute data control charts.

The methodology for calculating control limits for the four types of attribute data control charts is illustrated here. (See Table 5.3.) Keep in mind the definitions of defect and defective: *defects* are non-conforming items that are not necessarily defective (for example, a minor scratch on a used car), whereas a *defective* item is one that fails to meet the customer acceptance criteria regardless of the number of defects (for example a missing engine on a used car). For attribute control charts we will target a *random* sample size of 15 to 20 units per sample and at least 10 to 15 samples or subgroups (a sample size of 25 counts as one sample or subgroup) whenever possible (again dependent on the availability of solid data).

Table 5.3 Attribute control limit calculations.

	Upper control limit	Lower control limit	Center line
c chart	$\bar{c} + 3\sqrt{\bar{c}}$	$\bar{c} - 3\sqrt{\bar{c}}$	$\bar{c}$ = (SUM of all defects/ # of subgroups)
u chart	$\bar{u} + 3\sqrt{(\bar{u}/n)}$	$\bar{u} - 3\sqrt{(\bar{u}/n)}$	$\bar{u}$ = (SUM of all defects/ SUM of sample size within each subgroup)
np chart	$n\bar{p} + 3\sqrt{(n\bar{p}(1-\bar{p}))}$	$n\bar{p} - 3\sqrt{(n\bar{p}(1-\bar{p}))}$	$n\bar{p}$ = (SUM of all defective units/ # of subgroups)
p chart	$\bar{p} + 3\sqrt{((\bar{p}(1-\bar{p}))/n)}$	$\bar{p} - 3\sqrt{((\bar{p}(1-\bar{p}))/n)}$	$\bar{p}$ = (SUM of all defective units/ SUM of sample size within each subgroup)

Calculating control charts for continuous data requires a more complex methodology. To ensure you are learning tools that can be used to make an immediate impact, we will focus on the most commonly used continuous data control charts, the X-bar (average) and R chart (range). (See Table 5.4.) For this type of control chart we will target a constant sample size of approximately three to seven for each subgroup (sample size should never be less than two or more than nine). This type of control chart has several advantages:

- It will highlight changes to the average between subgroups (visually display changes to the process central tendency).
- It will detect changes within the subgroup (visually display the precision of the process).
- It is more sensitive to process shifts than some of the other continuous data control charts.

Table 5.4 X-bar and R control limit calculations.

	Upper control limit	Lower control limit	Center line
$\bar{X}$ and R	$\bar{\bar{X}} + A_2\bar{R}$	$\bar{\bar{X}} - A_2\bar{R}$	$\bar{\bar{X}} = (\bar{X}_1 + \bar{X}_2 + \bar{X}_3...)/k$
	$D_4\bar{R}$	$D_3\bar{R}$	$\bar{R} = (R_1 + R_2 + R_3...)/k$
			k = the number of samples or subgroups

In these calculations, A and D represent predetermined constants based on sample size. In a sample size of five, A_2 would equal .577, D_3 would equal 0, and D_4 would equal 2.114 (see Table 5.5).

Table 5.5 X-bar and R control chart constants.

Sample	A_2	D_4	D_3
2	1.880	3.267	0
3	1.023	2.574	0
4	0.729	2.282	0
5	0.577	2.114	0
6	0.483	2.004	0
7	0.419	1.924	0.076
8	0.373	1.864	0.136
9	0.337	1.816	0.184

So how do you know when your process is out of control? First, you must understand the difference between common cause and special cause variation. Common cause variation is inherent in the process (the design, machines, or other factors chosen as part of the system). Special cause variation is not inherent to the process and can be in large part eliminated or reduced (for example, human errors). Special cause variation is what we target when using control charts; this type of variation can lead to assignable root causes that can be reduced or eliminated to improve the process. Before we can identify special cause variation, we must first understand the various zones on a control chart. (See Figure 5.8.)

Control charts are valuable because they permit quick interpretation of special cause variation. Occurrences that should flag the control chart for further investigation include the following:

- A single data point outside of the control limits
- 9 consecutive data points on one side of the control chart center line
- 6 consecutive data points either increasing or decreasing
- 2 out of 3 data points on the same side in zone A
- 4 out of 5 data points on the same side in zone B
- 14 consecutive data points that alternate between up and down
- 15 consecutive data points within zone C
- 8 consecutive data points on either side of the center line with 0 in zone C

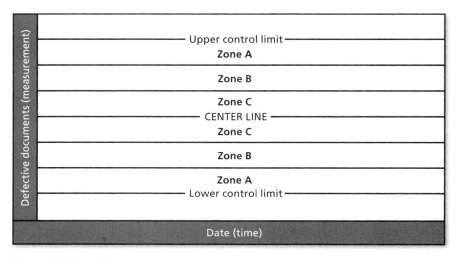

Figure 5.8 Control chart zones.

If one of these out-of-control items is true, then it is important to ask these questions with your LSS team:

- Have there been significant changes to the measurement plan or system (for example, new measurement tools or approaches)?
- Has personnel turnover resulted in someone new providing the measurements?
- Have there been significant changes to the primary inputs to the process (for example, personnel turnover, new supplier, new machine, seasonal weather impacts, or changes in guidance, procedures, and training)?

If the response to any of these questions is "yes," the issues should be further investigated for special cause variation. We have already discussed some tools that can help to eliminate variation in the process quickly (for example, standard operating procedures).

· Because control charts are an instrumental tool to any LSS program, it's useful to follow a theoretical example for creating a control chart and identifying special cause variation:

- The first step is to collect the samples, determine X and Y axes, and plot the information on a chart (same approach as a run chart).
- Next, calculate the control limits based upon the data type (attribute or continuous) and the sample size.

- Finally, calculate the various zones. Zone C is +/-1 standard deviation from the center line, zone B is +/-2 standard deviations from the center line, and zone C is +/-3 standard deviations from the center line. Figure 5.9 illustrates this.

Are there any out-of-control indicators on the example? Absolutely. (See Figure 5.10.) When you identify the first out of control data point, stop the process and investigate further for special cause variation by responding to the questions provided earlier.

If your process is in control and there is no special cause variation, will you meet the needs of the customer every time? Maybe not. Having a process that is in control does not guarantee that you are meeting the CTQs or specification limits. It simply means that you have a robust process that is in control. Control charts will not solve all of your problems, which is why the LSS methodology is often referred to as a toolkit. The primary goal of control charts is to help identify variation and sustain improvements. Control charts are one of the more complex tools in the LSS toolkit; combined with others, they are essential for sustaining not only a single LSS project but also the entire LSS program.

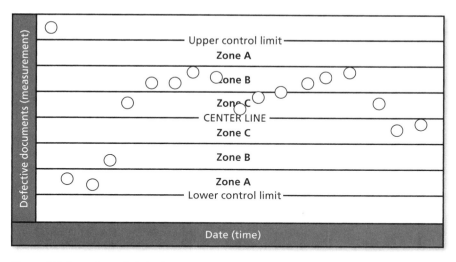

Figure 5.9 Example control chart.

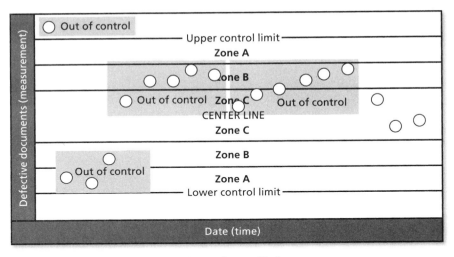

Figure 5.10 Example control chart with out-of-control indicators.

HISTOGRAMS

Impact in Overcoming Public Sector Challenges								
Breaking down stovepipes	Creating urgency	Leadership support	Metrics	Common goals	Increasing customer focus	Reducing impact of turnover	Overcoming complexity	Fostering collaboration
	X		X				X	

A histogram is simply a bar chart of specific data you have collected. The example described here relates to the measurement of wait time, which is represented on the X-axis, and the number of occurrences of each specific amount of wait time, represented on the Y-axis. A sample size of 30-50 data points is the minimum number required for a robust histogram. With this sample size, the histogram will allow you to answer a number of questions:

- Is there a tendency for the data to congregate around a specific range (to identify any central tendency)?
- Are there outliers in the data or a lack of central tendency (that is, does variation exist in the process)?
- Is the overall distribution normal?

The histogram is a simple but powerful tool. To effectively identify possible improvement areas, you must first understand the primary types of distribution.

The first is a normal distribution. Figure 5.11 displays a normal distribution that has a clear central tendency with a wait time between seven and nine minutes. If we were to review our specifications for this process we could determine whether it meets the requirements of the customer. If the distribution were equal or close to being flat across all bars, then we could assume there is significant variation. This is also an indicator of a non-normal distribution, which means either the data must be further stratified or the central limit theorem must be applied in order to help normalize the data (a normal distribution is required for many statistical tools).

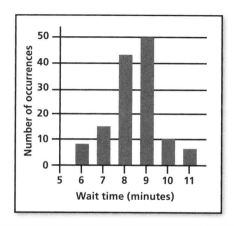

Figure 5.11 Example normal distribution histogram.

In the next figure we see distributions that may be non-normal. Figure 5.12 demonstrates skew either to the left or right. Where the distribution seems to be non-normal, the LSS team should investigate the process for out-of-control indicators (as discussed in the control chart section) or research the validity of the measurement plan/system.

Figure 5.13 demonstrates a non-normal bi-modal distribution (there are two peaks) and a general non-normal distribution example. These processes should be further investigated for out-of-control opportunities or variation.

134 Section Five

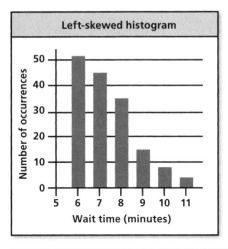 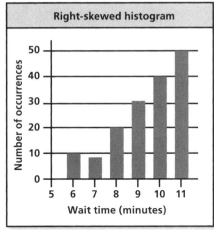

Figure 5.12 Example non-normal distributions.

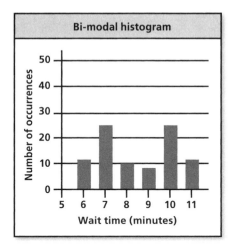 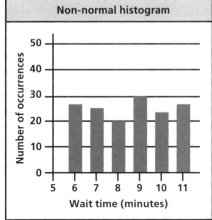

Figure 5.13 Additional non-normal distributions.

PARETO CHARTS

Impact in Overcoming Public Sector Challenges									
Breaking down stovepipes	Creating urgency	Leadership support	Metrics	Common goals	Increasing customer focus	Reducing impact of turnover	Overcoming complexity	Fostering collaboration
	X		X				X	

Pareto charts are one of the seven basic tools of quality. They are based on work initially completed by Vilfredo Pareto, an Italian industrialist and philosopher. Pareto observed that 80% of the property in Italy was owned by 20% of the population. Eventually, this would become the fundamental principle of Pareto charts, where 80% of defects occur due to 20% of possible causes. Although a very powerful tool, Pareto charts are simply a sorted bar chart with the highest occurring items on the left and a cumulative sum percentage line sketched across the top of the bar chart. See Figure 5.14.

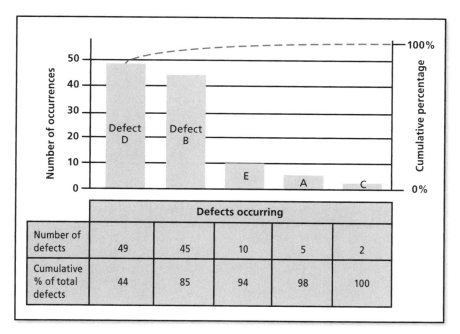

Figure 5.14 Example Pareto chart.

Pareto charts allow an organization to focus resources on the highest defect areas first. These charts are especially useful for determining an initial set of projects. For example, the Pareto chart illustrated in Figure 5.14 would indicate the need to focus resources on identifying a solution for defect D and defect B, which make up 85% of the total number of defects.

One of the fundamental differences between a histogram and a Pareto chart is the type of information obtained. With a histogram only information about the frequency of an item occurring is displayed (for example, errors, defects). On the other hand, a Pareto Chart clearly displays both the frequency of occurrence and the largest areas for improvement.

When creating a Pareto chart, the first step is to collect the proper data. Pareto charts typically display the number of occurrences on the Y-axis; and the types of occurrences are displayed on the X-axis. Once the data is collected, the next step is to create a sorted bar chart; the most frequent occurrence will display on the left side and the least frequent on the right. A cumulative total can then be calculated by summing across the columns and dividing that number by the total number for the Y-axis. For example, the total number of defects is 111 and defects D, B, and E total 104 occurrences. The total cumulative total at defect E is 104/111 or 94%.

So what if defect D is too large an opportunity, in terms of scope, to overcome with one project. This is where data stratification or "digging into defects" begins. To further stratify defect D, you will create Pareto charts with information on defect D only, specific to how, when, or why the defects are occurring: by defect type (incomplete, too hot, too cold, or other critical-to-quality elements), by location (different facilities), or by shift, operator, or resource. This is not an all-inclusive list, but the CTQs you determined should help you drive into the defect types.

Similar to the fishbone diagram, Pareto charts can be used to further drive to a potential root cause (this is also called data stratification). Figure 5.15 illustrates how data stratification can be used to further define a cause-and-effect relationship. In this example we see that defect D, when further investigated, is primarily caused by miscoded documentation.

Pareto charts are a valuable tool for analyzing specific opportunities and also for identifying initial or future projects. The key is to continue driving or slicing Pareto charts across different defect types or areas (such as location, operator, CTQs, or time-scale) to identify true high defect areas.

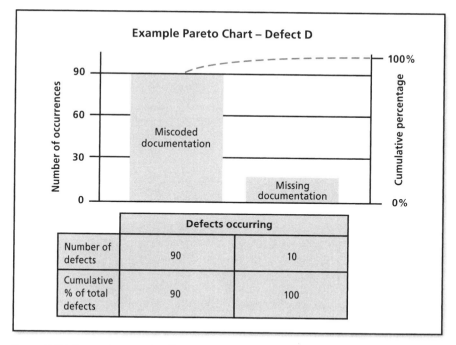

Figure 5.15 Pareto chart used to illustrate data stratification.

Public Sector Project Highlight

Pareto Charts

As part of a recent LSS initiative, the LSS MBB scoped an initiative from a data set involving 15,000 disparate defects using multiple sets of Pareto charts and data stratification to translate those 15,000 defects into a set of six high-value initiatives. Although completing the analysis took several months, the ROI definitely offset the resources utilized. This not only helped scope the initiative into bite-size chunks, but also helped identify an 80% solution that provided for millions of dollars in cost avoidance per year.

SCATTER DIAGRAMS

Impact in Overcoming Public Sector Challenges								
Breaking down stovepipes	Creating urgency	Leadership support	Metrics	Common goals	Increasing customer focus	Reducing impact of turnover	Overcoming complexity	Fostering collaboration
	X		X				X	

Scatter diagrams assist in identifying correlation between two discrete or continuous variables. *Correlation* is defined as the mutual (positive) or reciprocal (negative) relationship between two variables. The easiest way to describe the difference between a positive correlation and a negative correlation is to provide two scatter diagrams as an example (note that all correlation examples provided in this section are for example purposes only and are not based on actual data).

In Figure 5.16, the two variables exhibit a positive correlation. This means that as one variable increases, the other variable increases. It can also mean that as one variable decreases, the other variables decreases. The positive correlation demonstrated in this example is between an increase in the sales of marshmallows and skiing accidents (this may seem like an extreme example, but it will be used later to show a key caveat to scatter diagrams).

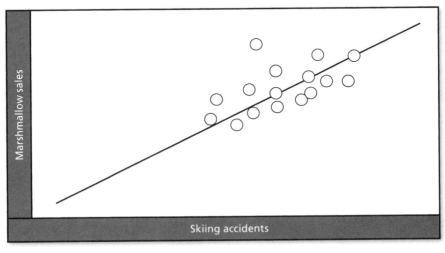

Figure 5.16 Scatter diagram exhibiting a positive correlation.

Figure 5.17 demonstrates a negative correlation between two variables. When two variables are negatively correlated, one variable increases as the other variable decreases. Similarly, when one variable decreases, the other variable increases. The example provided displays that the fuel efficiency for a vehicle decreases as the towing capacity increases.

There is one other potential outcome for a scatter diagram: the demonstration of non-linear correlation (illustrated in Figure 5.18). When this type of outcome is generated, it is assumed that the correlation

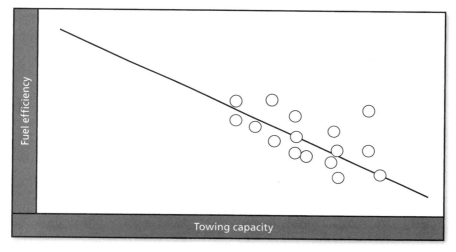

Figure 5.17 Scatter diagram exhibiting a negative correlation.

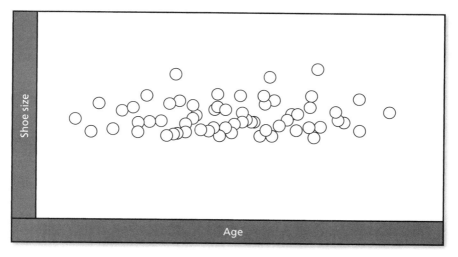

Figure 5.18 Scatter diagram exhibiting a non-linear correlation.

between the two variables is either not strong (neither positive nor negative) or that it cannot be defined. In this instance, other tools must be used or additional data collected in order to identify areas for improvement. Non-linear correlation can take many forms in terms of a scatter diagram (straight line across, u-shaped, and so on), but the one assumption can be made: If it is neither a positive nor a negative correlation, then it must be a non-linear correlation.

Back to our positive correlation example: If we stop eating marshmallows, then we will greatly reduce the risk of having a skiing accident, correct? Absolutely not; scatter diagrams demonstrate a correlation between two variables that should be investigated further using other LSS tools (such as cause-and-effect diagrams, Pareto charts, histograms, and so on). The output from a scatter diagram should never be the only rationale for changing a system or process (it does not identify whether true causation exists).

Let's take the marshmallow and skiing example further. Upon further investigation (using a 5 Whys approach) the team may determine that people purchase more marshmallows during the ski season due to an increase in hot chocolate sales. People tend to purchase more hot chocolate in winter when the weather is colder. Finally, during the winter season, there is a high likelihood of ice; these icy conditions may be the true cause of the skiing accidents. We see that it is not the marshmallows that cause skiing accidents but rather the potential for icy conditions.

It is important to note that other statistical tools can be used to investigate correlation in further detail (such as regression analysis). For the purposes of making an immediate impact, scatter diagrams provide a powerful first look into correlation and may provide insights into potential improvements.

Scatter diagrams are built using the follow steps:

1. Identify the variables you would like to investigate. This should be part of a robust measurement plan that describes the sample size, who will collect the data, the format for data collection, and the process for collecting.

2. Create a simple plot graph showing data points for the two variables.

3. Linear correlation or regression analysis tools provide an output with a specific correlation line. In place of these, use the input on your plot graph to draw a line based on the visible correlation of the data. This is not a perfect science, but it can provide valuable insights and a better understanding of correlation without a need to learn the more advanced tools.

4. Interpret the line to identify which type of correlation exists using the key illustrated in Figure 5.19 (if it does not fit any of these four generic categories, then it should be assumed that the scatter diagram is non-linear).

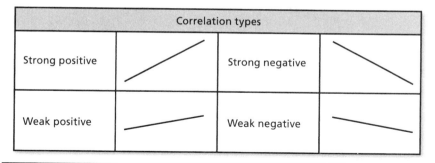

Figure 5.19 Correlation types.

Some items to keep in mind when interpreting the output from your scatter diagram:

- The stronger the relationship, the higher the likelihood that one variable has a direct impact on the other (further investigation is always required to avoid a marshmallow/skiing accident relationship).
- Scatter diagrams can be used to investigate the impact of multiple input variables: "Xs" (for example, time, operator) on one axis and "Ys" (for example, defects) on the other.
- Keep in mind that f(x) = y. The output gained from a process (y) is always a function of the quality of the inputs (x). All inputs as identified in your SIPOC or current state process map should be investigated for both correlation and cause.

IMPLEMENTATION PLANS

Impact in Overcoming Public Sector Challenges								
Breaking down stovepipes	Creating urgency	Leadership support	Metrics	Common goals	Increasing customer focus	Reducing impact of turnover	Overcoming complexity	Fostering collaboration
X	X	X		X		X	X	X

When improvements have been identified and the project is considered a success, there is nothing left to do but celebrate, right? An often overlooked aspect to every LSS project is the need to create a solid implementation plan. Many challenges must be overcome as part of implementing sustainable improvements, and one of the most critical is change management. At this point, if you are following the approach discussed throughout this book, you have already started to gain significant buy-in and support for your improvements (as part of defining the project, collecting data, analyzing the data as a group, and facilitating improvement events).

For each improvement, there should be a specific communication strategy and plan that informs all affected stakeholders of what changes will be made, when and how changes will affect their jobs, and whether there will be training or other communication relative to process changes. This helps to ensure that everyone is aware of improvements that will be implemented and how those improvements will directly impact their day-to-day job functions. Below are the critical inputs for a robust implementation plan that can be tailored for your specific organization:

- A prioritized list of all improvements to be implemented.
- A detailed, step-by-step approach to implementing the improvements.
- Categorization of each implementation step (for example, guidance changes, law or policy changes, training requirements, process changes, or standardization and system updates).
- Communication plan for implementing each improvement.
- Assignment of ownership for each detailed implementation step.
- Creation of a high-level time line for each implementation step (anticipated start and end dates). This can then be rolled up for each improvement, as well as the overall improvement plan to be communicated as the implementation time line to the executive steering committee.
- List of resources required from the executive steering committee (for example, leadership communication, funding).
- The final recommended aspect of an implementation plan, when linking it to the overarching improvements, is to include the anticipated value of the improvement (cost savings, defect reduction, and so on), in order to gain sign-off from the executive steering committee. There should also be a clear linkage to the initial project charter, Vo"X", value analysis, histogram, Pareto chart, or other LSS tools that clearly identify the rationale for the improvement.

The key is to not only gain leadership support and resources to accomplish the implementation, but also to integrate key stakeholders in the development of the plan to assist in overcoming any change management challenges. Remember, if people are not involved in the process, they are less likely to buy in.

> **Public Sector Project Highlight**
>
> **Delays in Implementation**
>
> As part of the interview process, a common theme began to arise relative to implementation in the public sector: unnecessary delays. Most of these delays occurred due to a lack of stakeholder involvement throughout the initiative (leadership from parallel organizations or process owners) and the inability to reference a robust implementation plan. It is critical that LSS team members insert themselves as part of the implementation process and meet regularly to maintain momentum and visibility of the implementation plans. Another proven method for overcoming inertia relative to implementation is to integrate an implementation status for each project as part of the executive steering committee meetings.

SUSTAINMENT PLANS

Impact in Overcoming Public Sector Challenges								
Breaking down stovepipes	Creating urgency	Leadership support	Metrics	Common goals	Increasing customer focus	Reducing impact of turnover	Overcoming complexity	Fostering collaboration
X	X	X	X	X	X	X	X	X

There is nothing more detrimental to the success of a LSS program than a lack of sustainability. If the increased effectiveness and efficiencies gained as part of each LSS project are not sustained, you will begin to lose leadership support and lose overall momentum for the LSS program. All of that hard work may have been for nothing if there is not a robust sustainment plan.

Although sustainment plans are often an afterthought, one must be integrated as part of the overall project. This helps ensure that key stakeholders are involved in the development of the plan and that the executive steering committee signs off on it as part of the project close-out. A typical sustainment plan should include:

- Creating or updating standard operating procedures to reflect the changes to the current state process (including any training or checklist updates required).

- Creation of a transition plan from the LSS project team to the process owner.
- Validation of improvements gained as part of the improvement plan. This should link directly to the goal statement of the project charter.
- Updates to visible metrics and dashboards to integrate new or updated control charts.
- Integration of a continuous process improvement feedback loop to the process owner to ensure the improvements are sustained (for example, online feedback, organization idea programs).

SPOTLIGHT – PROS AND CONS IN UTILIZING EXISTING DATA

Using existing data is almost always faster, cheaper, and easier. Little stakeholder involvement and almost no senior leadership approval is required, but there are possible downsides.

The first item of concern is the timeliness and accuracy of the existing data. Did they use a robust measurement plan to gather the data or was it destined to reach a pre-determined conclusion? Has it been collected recently or was it for a previous data call? These are just some of the concerns relative to the timeliness and accuracy of existing data.

Another consideration is the continued need to gain stakeholder input and support throughout the project. This is an ongoing process, which is facilitated by involving the various stakeholders throughout the project. In this case, working with the stakeholders to obtain relevant data and possibly re-focus the initial measurement plan provides an opportunity for team members and other stakeholders to voice their concerns and recommend potential areas for improvement. This gives practitioners the ability to resolve concerns, answer questions, or provide general project clarification of the intent of a project well in advance of the implementation and sustainment of improvements.

Existing data can be used when the data is not critical to the overall project, but it is usually better to obtain new data. This ensures that the data used is relevant, timely, in the appropriate format, and accurate and it also provides the opportunity to communicate with the appropriate stakeholders, involve them in the project, and resolve concerns prior to the implementation of a project.

Section 6
Create a "Buzz"

"Government is slow to move, yet when once in motion, its momentum becomes irresistible."

– THOMAS JEFFERSON

The key to gaining and maintaining momentum relative to LSS in an organization is the sustainment of a constant "buzz." This buzz should be the culmination of results gained, awards and recognition provided to successful project team members, career development/enhancement opportunities, and the ability to build relationships outside of a team member's normal organization. It is the role of the project sponsor, team leader, and everyone involved in a LSS project to be communicating these key points both formally and informally every time an opportunity presents itself. If the stakeholders involved in a LSS initiative are not excited about the potential opportunities provided, it will be difficult to sustain the approach as a viable improvement methodology in the organization.

STAKEHOLDER INVOLVEMENT

Throughout this book we have discussed the importance of continued stakeholder involvement. It is critical not only to completing an initiative, but also to securing buy-in for the implementation of a specific effort and obtaining support for the overall LSS program. The key to obtaining and maintaining positive stakeholder engagement and support includes a number of factors:

- The active engagement of every stakeholder, at every opportunity. Even when specific stakeholders are engaged only to provide a certain data set as part of the measurement plan, they should feel as though they are active participants in the overall project. This can be completed by briefing them on the defined goals and objectives, as well as the anticipated outcome. Focusing on this as part of every interaction will help gain continued support.

- Inclusion of representative stakeholders (core project team members should be required) during the outbrief of the project. The outbrief should also include a high-level discussion of each team member's roles and responsibilities, any challenges they may have helped overcome, and kudos for the support they have provided throughout the project. Without every team member's input, no LSS program can succeed.
- Communication and recognition that include core team members and, if possible, anyone who impacted the outcome of the initiative (sponsors, team members, or parallel support). This displays that the success of any LSS initiative will be properly recognized, not only to the team leader and the LSS team leader but to anyone involved.

VISIBILITY

As part of using LSS tools and techniques, there is a continued focus on outcomes. This transfers extremely well in terms of visibility. As projects are completed, they should be included as part of an organization's dashboard. The information displayed should include the results gained (for example, 25% increase in customer satisfaction, 55% decrease in cycle time), names of project team members involved and their roles, any additional improvement projects identified as part of completing the initiative, and most importantly, what the savings allowed you to do instead (buy 100 necessary computer upgrades, provide for an additional .5% increase in the salary pool). This translates into visibility for the improvements, the team members, other potential improvement initiatives, and the power of LSS.

CONTINUED COMMUNICATION IS CRITICAL

If you are not communicating the successes of your LSS initiatives, no one will understand the power of the methodology. If no one understands the power of the LSS methods, the program will fail. Communication of project results is the single most important aspect of gaining and sustaining momentum for your LSS program. It is through demonstrated and sustainable results that personnel will begin to proactively become involved and create a buzz across the enterprise. It is the responsibility of the LSS team leader to leverage every opportunity to discuss results. The primary opportunities include briefing senior leadership, creating an enterprise-wide newsletter highlighting teams and results, creating an award for the best project completed in each quarter, and developing an intranet page providing highlights on every project completed. This is not meant to be an all-inclusive list, but it should provide a starting point for a robust communication strategy.

The other critical but often overlooked item required for a sound communication strategy includes the development of a LSS elevator pitch. In 60 to 90 seconds, you must be able to describe the value the methodology provides (a systematic, data-driven approach to continuous process improvement focusing on providing customer value and eliminating waste) and examples of results specific to your organization (eliminated 800 hours of processing and cycle time from our brownie example). This will help the perceived value permeate the organization.

NO SUCH THING AS A PROBLEM, JUST OPPORTUNITIES

As part of fostering a non-attribute environment across the organization, it must be clearly communicated that there is no such thing as a problem; there are only opportunities. The tone of the leadership, project team leads, team members, and anyone else involved in the LSS journey must focus on creating an organization of excellence thereby removing barriers created by legacy processes, relationships, or ideas. Practice using this tagline: *There is no such thing as a problem, only opportunities to better the organization.* The caveat here is ensuring that leadership continues to strive for this in all of their interactions across the organization and "live it" as role models within their own functional areas.

AWARDS AND RECOGNITION IN THE PUBLIC SECTOR

Recognizing project team members for a job well done is an often overlooked step in the LSS process. Without the hard work, dedication, and inputs from multiple stakeholders across an organization, most projects would be unsuccessful. The time, commitment, and desire to improve demonstrated by successful LSS teams should be communicated, rewarded, and recognized across the entire organization. It may be difficult to provide the typical private sector rewards in a public sector environment (immediate bonuses, extra time off, or team dinners), but there are ways to reward success in the public space. Recognition can take the form of communicating the success to senior leadership, which in turn increases visibility for individual team members. The team can be recognized in small ways such as providing organizational coins, pizza parties, or even the ability to present their successes and LSS story to the directors of an organization. Even though some of the normal recognition methods in the private sector cannot be used (monetary rewards), an appropriate reward must be determined to recognize the team's success. Without this, the success of a LSS program can be significantly jeopardized.

> **Public Sector Project Highlight**
>
> **Recognizing All Project Participants**
>
> During the completion of a recent public sector LSS initiative, it became apparent the only people receiving visibility for the results were the LSS belts. They were providing briefings up the leadership chain, communicating the success of the project, and including the results achieved as part of their annual reviews. In order to gain momentum and support from across the entire enterprise, *all* project participants must be recognized for their time, dedication, and commitment to process improvement. The easiest way to overcome this is to have the entire team brief the results. If some team members are unable to attend in person, a conference call will suffice. Just knowing you had an impact and being able to communicate that to leadership will create excitement for the next project.

SPOTLIGHT – IMPORTANCE OF SOFT SKILLS IN LEAN-SIX SIGMA

LSS is a systematic and data-driven approach to process improvement. This does not mean that there is not a relationship or human capital management aspect. Having the appropriate soft skills to maintain support can be the difference between a successful LSS program or project and a failure.

Gaining and maintaining support requires empathy. Supporting the LSS program is most likely not the primary job responsibility for most of the project stakeholders. For example, if there is a need to extend the time line by a couple days in order for someone to complete an aspect of the measurement plan, make the extension and update the time lines as appropriate. The reason a majority of the basic quality tools are so valuable is because they allow stakeholders to voice their opinions and become part of the process. It is the role of the LSS team leader to ensure a continued focus on relationship management, change management, communication, listening to everyone's opinion, and fostering a non-attribute environment while respecting all stakeholders.

Section 7
Sustainment

"Quality is not an act, it is a habit."

– ARISTOTLE

There are four key elements to successfully sustaining any LSS initiative: assigning a process owner, creating a robust sustainment plan for the process going forward (standard operating procedures), implementing visible metrics, and providing a mechanism to foster continuous process improvement. Although we covered these four critical areas as they relate to sustainment earlier in this book, there are other areas involved with sustaining a new process that we have not discussed in extensive detail.

DEALING WITH TURNOVER

Resource turnover is seen as a given in the public sector. This can be caused by military deployments, administration changes, job moves due to promotion, and a variety of other factors. This is where the fundamentals of LSS can assist in sustaining an environment of continuous improvement. Everything that you have completed throughout the initial deployment of LSS (for example, executive, belt, and awareness training), as well as the documentation completed throughout each initiative (for example, project charters, process maps, and improvement plans) documents why the project is being pursued, what you believe the possible outcomes may be, what resources are required, and how you plan to implement and sustain potential improvements. The key is to document the use of each tool, the outcome that was reached, and which stakeholders were involved. It is also critical to facilitate the tollgate reviews, whenever possible, to ensure a continued focus on LSS efforts by leadership. Finally, one of the most beneficial items for overcoming turnover is the sustainment of an executive steering committee. This helps to ensure that a single person does not drive the LSS program, but rather a board of executive sponsors who are tied to specific positions that are not affected by turnover.

PROJECT AND DATA TRANSPARENCY

One of the ongoing messages from the executive steering committee must be focused on complete project and data transparency.* The key areas in which data plays a critical role include: project selection, project execution, and communication of project results. During the project selection process, leadership must clearly articulate the message that all projects identified should be completed with a non-attribute environment in mind. What that means is that people who identify a defect or opportunity for improvement will not be penalized for being part of the process (don't kill the messenger). This will help ensure that all possible improvements are identified without a fear of retribution from leadership.

The LSS approach has two common themes: systematic and data driven. LSS projects cannot be successful without access to data. It is the responsibility of the project lead to identify data requirements as part of the measurement plan, but it is the responsibility of leadership to ensure everyone is committed to providing this data. Leadership should be communicating what data is required and why it is required in order to create an environment of teaming across the organization.

The final requirement for project and data transparency is sometimes the most difficult: being open and clear about results gained from a specific initiative, whether cost savings, defect reduction, or an increase in customer satisfaction. Leadership is sometimes unwilling to expose the fact that processes were not optimized in the past or communicate that they have identified potential savings. This may be due to an inherent fear that they will be perceived as previously poor performers, which may impact their ability to gain support and funding from public sector leadership (for example, the current administration, congress).

The need for project and data transparency is critical to the success of any LSS program. Use your funding reallocation plan to provide the rationale for needing critical data during the project selection and execution phases. Any results communicated should also be tied to the funding reallocation plan in order to offset a possible budget cut (cost savings were immediately re-infused into a critical mission requirement). This will help gain support from senior leadership and help support an environment of continuous process improvement.

---

* One caveat to project and data transparency is public sector security requirements (for example, classified, secret, or top-secret information). If your project has such requirements, work with your leadership and security contact to ensure the classification of information is never jeopardized.

CIVILIAN COMPONENT

As we discussed earlier, frequent turnover sometimes seen in the public sector can have a significant impact on sustaining a LSS program across the enterprise. Out of all the various employee types (for example, military, civilian, and contractor), the civilian component typically has the lowest turnover rate. It is beneficial to leverage this component for continuity of operations. Ensure that civilians are offered appropriate access to the senior leadership steering committee meetings and training and involve them in a majority of the LSS efforts. This will help the organization not only sustain a knowledge base and leadership buy-in, but will also foster an environment of continuous process improvement. People who participate in successful LSS programs create a buzz about the savings they have seen using the methodology. New employees will notice.

IMPORTANCE OF GOOD GOVERNANCE

A solid governance structure is instrumental to the success of any LSS program (see Figure 7.1). As discussed previously, the first step in creating a robust governance model is the identification, staffing, and sustainment of an executive steering committee. Once this committee has been identified, it is their primary objective to ensure a solid pipeline of potential LSS initiatives, a visible LSS balanced scorecard, and constant communication about project successes both internally and externally to the organization. They must also ensure that candidates are identified for

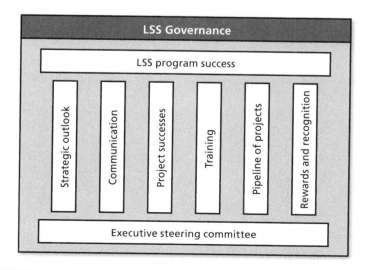

Figure 7.1 Robust governance structure.

LSS training opportunities, that rewards and recognition are provided as each project is successfully completed (regardless of the amount of savings), that projects are tied to the strategic plan as defined by the executive steering committee and other representative stakeholders, and that the appropriate support is provided from both a resource and momentum perspective to sustain the LSS program. Without a solid governance structure in place, with an executive steering committee as the foundation, any LSS program is likely to be unsuccessful, regardless of whether it is in the private or public sector.

CAPTURING LESSONS LEARNED

I have never completed an improvement project in which I did not learn how to approach the problem from a different angle, better engage stakeholders, more accurately define the measurement plan, or otherwise identify an opportunity to improve the manner in which I facilitated an initiative. This is part of the learning process and every LSS effort is a learning opportunity. The hurdle that must be overcome in regard to communicating lessons learned is communicating them across the entire enterprise. Although this may seem like a daunting task, it can actually be accomplished with relative simplicity. As you are creating a repository for future projects, keep in mind the need to also create a repository of lessons learned. The information included as part of this repository should include the original project charter, the type of project completed (manufacturing, transactional), the team leader for the project, when it was completed, what the results were, the obstacles faced or the lessons learned, and any approaches developed or used to overcome potential roadblocks. With relative ease, this extremely valuable information will save your LSS program time, money, and resources. Who wants to make the same mistake more than once when a solution has already been developed? All it takes is the initial foresight to build the repository on the front end.

REPOSITORY OF BEST PRACTICES

There is nothing worse than improving a process, product, or service within an organization and knowing it is not being deployed across the board. This is where a repository of best practice approaches can support enterprise-wide deployment. The information maintained within the best practice repository should include the current process owner with relevant contact information, a finalized process map, standardized operating procedures or checklists, any lessons learned, and other relevant information or data. This data can be stored in an internal location such as an easily accessible computer folder. If possible, it should be deployed via an intranet to allow for global access. This way the best practices are not only being deployed across the organization, but the process owners are also being credited with improving their respective parts of the organization.

> **Public Sector Project Highlight**
>
> **Sharing Best Practices and Lessons Learned**
>
> It is common to share only process improvement best practices and lessons learned as part of the LSS program, but that is not optimal. As part of each improvement, be sure to document best practices or lessons learned relative to systems and people as well (remember the pyramid of processes, systems, and people). This will ensure that approaches to leverage information technology, organizational design, or other areas are included as part of the best practice and lessons learned repository.

REPOSITORY OF FUTURE OPPORTUNITIES

Another common pitfall to utilizing LSS in the public sector includes not having a robust repository of future opportunities. As the executive steering committee is tracking overall progress, resources are completing training, or teams are completing improvement initiatives, they must be able to tap into a repository of potential projects in order maintain momentum. Ideally this repository will be located in an environment that is accessible to everyone across the internal organization (intranet, shared drive) and contain some basic information for selection and sorting.

The basic information required as part of this repository should include:

- The proposed "owning" organization (who is the current process owner) and the person who identified the initial opportunity (in case further clarification is required),
- The date the project was identified and a proposed revisit date (to keep the repository up-to-date),
- The project charter and prioritization information that was collected as part of the project selection meeting (opportunity and goal statement, stakeholders impacted), and
- Linkages to the current strategic plan (if the strategic plan has changed since this project was identified, is it still a viable potential opportunity).

Without a repository of solid future opportunities, it is difficult to sustain excitement and the overall LSS program. This repository should also be used as a method for employees to propose potential projects. Allowing everyone to be involved as part of the LSS program helps to ingrain the tools, methods, and desire to improve into the overall organization.

AGILITY

The public sector can sometimes differ with the private sector in terms of the level of leadership support. As evidenced by the independent interviews facilitated as part of this book, it was discovered that a majority of the power to influence lies with middle managers. This is why buy-in must be sought at this level and then leveraged in leadership meetings (bottom-up approach), informal middle management discussions (across approach), and formal project meetings (top-down approach). Once this buy-in has been gained and sustained, it will be the responsibility of both the project leaders and middle managers to remain resilient. This includes tailoring tools, methods, approaches, and briefings to the specific leadership preferences. It will also require that simulations and training be tailored to various environments. Ultimately, the goal is to leverage the power of the LSS tools. If the LSS program is not agile enough to tailor the approaches, then it will likely fail.

USING LEAN-SIX SIGMA FOR ORGANIZATIONAL DESIGN

Members of an organization once voiced concern that they were not achieving results from their monthly status meetings. It seemed as though each person was focusing on the "fire of the day" or what they believed was important to them. There was not a focus on how work was being completed on a daily basis and the overall impact on the organization's mission and vision. There was no formalized structure to understand how each resource impacted the overall strategic plan. It was necessary that the organization identify an approach to prevent everyone from "talking around the table." Fortunately, LSS tools, techniques, and methods can be adapted for a variety of purposes. If an organization is trying to decide whether a re-organization is required or determining the most robust structure for their organization, the LSS approach can be instrumental in optimizing the final outcome.

The process for using LSS to revitalize an organization begins with defining the strategic plan, mission, vision, SMART goals, SMART objectives, and SMART metrics. Once this has been completed, a roles and responsibilities matrix, position descriptions (constructed using affinity diagramming and input from the roles and responsibilities matrix), and resource names can be mapped to the above items. See Figure 7.2.

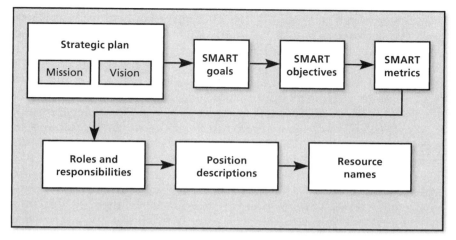

Figure 7.2 LSS for organizational design.

This in turn ensures that every resource within an organization has a direct impact on the strategic mission and vision, which typically has a positive effect on overall morale. It also provides a framework and agenda for each status meeting. The key is to involve all members of an organization in the process. This will provide the foundation for a robust change management approach and implementation strategy and allow personnel to proactively voice their concerns in advance of implementation. The purpose of this is not only to optimize the ability to achieve an organization's' strategic plan, but also to increase the ability for employees to see their relationship to bottom-line results, making performance-based evaluations much easier using quantitative measures as the foundation for promotions or other career progression opportunities. Sometimes it's necessary that meetings focus on items outside of the strategic plan, but if this becomes a recurring theme, then it may be valuable to revamp or revitalize an organization's strategic plan or its organizational structure.

SPOTLIGHT – PERFORMANCE MEASUREMENT

We have already discussed the importance of visible, balanced scorecards that are a crucial part of any organizational performance measurement strategy, but this topic is so critical to the sustainment of LSS or any organizational improvement method that it merits further discussion. The use of data must become the underlying principle for all long-term and short-term decision making. This is not possible without a robust performance measurement system.

As we discussed in the scorecard section, any data collected and continuously reviewed by an organization should somehow be tied to the overarching strategic plan. If the strategic plan was appropriately deployed, everyone should be aware why the goals, objectives, and metrics are critical in achieving the mission and vision. To ensure continuity, data must be easily gathered and interpreted. This requires a standardized definition for collection and interpretation of the data whether automated or manual (similar to a robust measurement plan). The time frame for reviewing the data should be set by leadership, with input from the organization in order to proactively identify and react to trends in performance. The easiest approach to identifying these trends is by establishing thresholds for a "red," "yellow," or "green" status. It is important that each performance measure also have an ongoing response as to why the metric is within that specific threshold. Even green metrics should have a rationale for achieving this mark, as other organizations may be performing in either the yellow or red status and a best practice may be available to benchmark across the organization. The operational definition and links to the strategic plan must also be part of the performance measurement input (the *who, what, when, why,* and *how* of measurement).

Performance measurement must be seen as an opportunity to evaluate how the organization is doing relative to its strategic plan, identify areas requiring improvement, and motivate continuous process improvement through data and leadership visibility (if a red metric is identified, an environment of collaboration and support should help identify a team, resources, and a plan for improvement), and celebrate success (there must be appropriate recognition when a green metric is obtained and sustained). I think it is important to repeat one of the items above: the identification of potential improvements should always be seen as an opportunity. If an environment of collaborative, continuous process improvement is not fostered, organizations are likely to "game" the performance measurement system.

Section 8
Conclusion

*"The quality of an organization can never
exceed the quality of the minds that make it up."*

– DWIGHT D. EISENHOWER

The LSS approach must be tailored slightly for the public sector, just as it was for non-auto manufacturing industries, services industries, and back office operations. The public sector is ripe for improvement and the leadership is dedicated to making sure improvements occur. I have had the opportunity to work with some great project sponsors, leaders, and executive steering committees in the public sector. They had a single common theme: the drive to improve. This is the fundamental criteria that must permeate up, down, and across the organization. If this is achieved, the tools, results, and sustained improvements will follow.

The keys to the public sector LSS approach include:

- Having a robust funding reallocation plan.
- Gaining buy-in from middle management, as they have a significant impact on the success of the program.
- Using LSS tools to overcome the challenges in the public sector.
- Creating an environment of excitement and fun in order to gain and sustain momentum. High impact initiatives, tied to strategic goals and initiatives, will help in accomplishing this.

HOW HAVE WE OVERCOME ALL THE PUBLIC SECTOR CHALLENGES?

As discussed in Section 2, there are many challenges to deploying and using LSS in the public sector. Some of these are not specific to the industry, but rather enhanced due to a multiplier effect. This is not to say that these challenges cannot be overcome. If the appropriate tools are used (as outlined in Sections 3–7) and momentum is gained and sustained, the challenges can be almost eliminated. In Section 2 and throughout the other sections, we overcame *all* of the challenges faced in the public sector environment using some of the most basic LSS tools. These LSS tools not only help overcome the obstacles of using LSS in the public sector, but also provide critical information about internal and external competitors (for example, SWOT analysis, benchmarking), trends (for example, run charts, control charts), and the correlation between multiple variables (for example, scatter diagrams, histograms). They provide high-value opportunities for improvement (for example, Pareto charts, project selection meetings), display potential causation (for example, fishbone diagrams), and help ensure sustainment (for example, control charts). We identified 22 tools to overcome stovepipes, 27 tools to help create a sense of urgency, 17 tools for gaining leadership support, 19 tools to establish metrics, 25 tools to help establish and drive toward common goals, 16 tools to assist in identifying and supporting a customer focus, 14 tools to reduce the impact of turnover, 26 tools to overcome complexity, and 27 tools to help foster an environment of openness and collaboration across the entire enterprise.

THERE WILL BE ITERATIONS ACROSS THE PHASES

One of the items that newcomers to the LSS methodology encounter is the belief that since the approach is meant to be systematic, there cannot be multiple iterations across the phases. This is almost impossible to avoid. For example, as practitioners complete Pareto analysis, they may realize additional data is required to validate potential improvements. As the SIPOC is being completed, the team may realize that a critical stakeholder is missing. The key in these and similar circumstances is to ensure the project champion and team leader are aware of the information update or new information requirement, sign-off is gained from the executive steering committee, and the necessary resources are provided to complete the change. If the team is not continuously involved, buy-in may not be achieved during implementation. A little additional work, communication, and buy-in throughout the project will payoff 100 times as the final improvement is being implemented and sustained.

OTHER PROCESS IMPROVEMENT METHODS

Keep in mind that there is more than one way to improve an organization. Although LSS is a powerful approach that can be tailored to almost any environment, it is not the only process improvement methodology available.

Some of the other common improvement approaches include:

- *Business process reengineering* – The easiest way to think of business process reengineering (BPR) is as a "big bang" approach. BPR is an approach used to fundamentally investigate and rethink the manner in which an organization completes processes in order to identify sometimes radical improvements. The standard phases in this approach include identifying and analyzing the as-is processes, developing a to-be process, and testing, implementing, and sustaining the to-be processes.

- *Business process management* – Business process management (BPM) is a holistic approach to managing organizational processes with a focus on delivering to client specifications, needs (CTQs), and wants (delighters). Similar to LSS, the drive is to optimize the enterprise using continuous process improvement tools. BPM typically follows a five-phase approach: designing, modeling, executing, monitoring, and optimizing.

- *Theory of constraints* – Introduced to the world in 1984 by Eliyahu Goldratt, this approach focuses on continuous process improvement by identifying the constraints in a process (the bottlenecks). Once these are identified, a systematic approach is used to remove them; the process is repeated to continuously improve.

- *TRIZ* – This approach was developed by Genrich Altshuller in 1964 as a way to develop new ideas or innovative solutions to existing problems. It focuses on identifying the problems and utilizing historical solutions, based on an algorithmic approach, to identify new system designs or changes to existing systems.

- *ISO Standards* – The International Organization for Standardization or ISO is a standard-setting body that was created in 1947. The purpose of ISO standards is to ensure an organization applies formalized business processes across the enterprise.

These are just the tip of the iceberg when it comes to other possible improvement methods. The key is to ensure you are using the right tool, approach, or method for the particular circumstances. Unfortunately there is no one-size-fits–all approach to process improvement. It is up to the project team to identify the appropriate approach and necessary tools (the simplest tools possible for a specific goal) and develop a collaborative, fact-based solution.

EVEN CHANGE LEADERS NEED TO CONTINUOUSLY ADAPT

Mass production, Deming's 14 points, statistical process control and concepts were all mind-boggling and ground breaking approaches when they were introduced. Similar to these approaches, LSS is a dynamic methodology that continues to transform itself to address different environments. Change leaders must be able and willing to continuously adapt. It is the responsibility of these leaders to research and identify new ways to optimize their organization.

As part of the interviews completed for this book, participants were asked what they thought was next on the organizational improvement horizon. The one recurring response was that although the names of the improvement approaches may change and other methods may be integrated, it would contain most of the tools used currently in LSS. The responses also referred to LSS as a "toolkit," an assembly of tools from a variety of previous methods and approaches. The value added from the methodology was the systematic approach. With this in mind, change leaders must remain vigilant and agile to address possible changes to the methodology. Although it may have a new name, it will likely be based on many of the same principles.

IMPORTANCE OF LEADERSHIP BUY-IN

We have already highlighted the need for leadership support to address the challenges in the public sector, as part of the executive steering committee discussion, and in many other sections of this book. So why highlight the topic again?

In completing the interviews for this book, a common theme became apparent about the middle manager, the other level of leadership support required in the public sector. Oftentimes leadership has a higher turnover rate in the public sector due to emerging requirements (we discussed how this is overcome through the executive steering committee). Another approach to overcoming this turnover is to gain support from the middle management ranks and organizational subject matter experts who may not have the title of "manager" but who have the ability to build excitement and buy-in across the enterprise.

How can this be achieved? By involving them in all aspects of the LSS program and asking, yes asking, for their opinions on the projects selected, the metrics identified as part of the strategic plan, and the implementation plans created as part of each improvement effort. This demonstrates that their opinions are valued and that they can leverage the success of LSS projects for their own gains (promotion potential, making a lasting impact on the organization, and long-term career fulfillment, to name a few).

Similar principles apply:

1. Lose the terminology where necessary.
2. Identify stakeholders with leverage to impact the success of the LSS program (not just leaders by title, but also leaders through expertise and other powers of influence).
3. Ensure everyone obtains LSS awareness training with tailored simulations to involve and excite them.
4. Allow them to provide input on project selection and charters.
5. Involve everyone as part of the strategic planning process.
6. Identify low-hanging fruit to make an immediate impact.
7. Reward and recognize individuals for project results and program successes.
8. Be able to clearly articulate the LSS elevator pitch. If you can't clearly articulate the value LSS provides to individuals and the organization, no one else will be able to do it either (remember WIIFM).
9. Work to integrate continuous process improvement as part of the culture.

Public Sector Project Highlight

Leadership Support

Some completed LSS projects are not rewarded with high praises. When deploying the implementation and sustainment plans for a recent LSS DMAIC project, the leadership team was not supportive of the improvements. It was not immediately apparent what the concerns where, but upon further discussion it became clear that the improvements identified did not link to the project charter. This meant that the implementation of these improvements had no impact on the strategic plan for the public sector organization. This can happen in any environment, but with the metrics for success being tailored for specific public sector entities (vs. profit or revenue for the private sector), it is critical that all improvements identified as part of the implementation plan be validated to ensure they meet the goals and objectives of the original project charter.

CONTINUOUS IMPROVEMENT

In order to have any chance for sustained success, an organization must foster an environment of continuous process improvement. This is an environment where people begin to identify opportunities for improvement as part of their job, instead of only during project selection meetings or when asked by senior leadership. An organization that is proactively looking for opportunities to improve will not only be at the forefront in terms of effectiveness and efficiency; it will be able to describe the steps they are taking on a monthly, weekly, and daily basis to achieve them (through goals, objectives, and metrics). There are many approaches to fostering an environment of continuous process improvement, but some key items are part of every successful LSS organization:

- Leverage technology to identify improvements through web-based dashboards, globally deployed standard operating procedures, and intranet "idea wells." This will demonstrate that improvements are a positive form of stakeholder input.

- Celebrate opportunities for improvement. This means fostering a non-attribute environment in every meeting, event, and discussion. If people feel uncomfortable or are threatened when identifying opportunities for improvement, it is likely that continuous process improvement efforts will fail.

- Continue to look for emerging improvement trends. There has been an increase in discussion of the potential use of radio-frequency identification or RFID chips to alert users of upcoming service requirements in cars, appliances, and other electronic equipment. This chip would automatically notify the user, as well as the manufacturer, if key components were about to fail. Remember, at some point LSS was an emerging improvement approach.

SPOTLIGHT – GREENING YOUR ORGANIZATION USING LEAN-SIX SIGMA

Similar to using LSS to improve process performance by continuously identifying efficiency and effectiveness opportunities, the LSS methodology must also continue to adapt to emerging trends and technology. One such emerging trend is the recent focus on a sustainable or "greener" organization. Although the focus is still on eliminating waste and variation from the enterprise, the approach must be tailored slightly.

If one of the primary goals is to create a more sustainable organization, this must be integrated into all aspects of the LSS program:

- As part of the executive steering committee, a chief sustainability officer or environmental oversight position should be assigned.
- The goals pertaining to greening an organization should also be integrated as part of the strategic plan, visible dashboard, and project selection criteria.
- When creating a robust measurement plan, additional focus should be provided to energy (for example, oil, coal, gasoline, electricity) and water savings. These savings can take the form of reducing requirements and reusing and/or recycling materials, energy, and water. The measurement data for most green initiatives are readily available. For example, this information is included as part of the cost of raw materials, facilities maintenance costs, energy, and water bills. If the organization is willing to take the approach a step further, it can begin to proactively become a more sustainable entity through advanced metering, deploying a green program management office, or creating a green strategic plan with a visible dashboard to track progress.
- Efficiencies in terms of green focus on using less, while effectiveness focuses on the better use of existing materials (for example, brown-field site construction).
- Defects are still a form of green waste. They include wasted materials, energy, and water.

LSS is a systematic approach to eliminating waste and variation from existing products or services. Similar to tailoring the approach for the public sector, the LSS tools can provide significant returns in creating more sustainable organizations. The current paradigm must shift in the organization, from focusing on the status quo to refocusing time, energy, and senior leadership on environmental sustainability.

Appendix A
Interview Questions

1. Total work experience (number of years):
2. Total LSS experience (number of years):
3. What has been your skills progression in the LSS methodology?
4. What are the top three differences you have seen in using Lean-Six Sigma in the public sector vs. the private sector?
5. What are some challenges you have faced implementing LSS in the public sector? What would you say are the top three challenges you consistently encounter?
6. What was your most challenging public sector engagement?
7. How have you overcome these challenges (using which LSS methods)? Why did you use this approach? What didn't work?
8. Rank the following challenges in order from biggest challenge to least challenging (1 = biggest challenge, 9 = least challenging).

    ____ Not typically focused on metrics such as profit or revenue generation

    ____ Lack of customer focus

    ____ Lack of common goals

    ____ Lack of leadership support

    ____ Hierarchical and stovepiped environments

    ____ Limited sense of urgency

    ____ Mix of various employee types

    ____ Constant employee turnover

    ____ High complexity of the public sector

9. Considering this list of challenges, where have you encountered them and how have you overcome them?
10. What do you believe is the most powerful Lean-Six Sigma tool (for example, strategic planning, VoC)? Why? Explain where you have used it and the impact.
11. What do you think the future holds for Lean-Six Sigma as a method? What do you see as future trends (for example, green, strategic planning)?
12. How would you describe perceptions of public sector projects before, during and after (yourself, your coworkers, management, and customers)?
13. How would you describe the change process related to the project in regard to the people factor (for example, coworkers, management, and customers)? What lessons did you learn that you'd share with others for these types of projects when it comes to managing change in a public sector environment?
14. Did you apply anything you learned during the project to other processes or areas? What have been the results of that?
15. Do you have any data that can be shared demonstrating results (for example, cycle time, defects, and financial benefits)? (All client-specific references will be stripped and the data will be converted into oblique references.) Beyond data, do you feel the project was successful? Why or why not?
16. What other thoughts or comments do you have on using LSS in the public sector?
17. Are there other individuals—coworkers, managers, or customers—who you think would be interested in contributing their perceptions to this book? If so, please supply name, company, position, and a point of contact.

Index

Page numbers in *italics* refer to tables and figures.

A

accountability, 16, 29
accounting and finance, 34
action plans, 1–2, 16–17, 114
affinity diagrams, 58
agendas
    effective meeting, 114
    executive steering committee, 19
    interview, *76*
    kaizen, *101*
    project kickoff, *66*
    project selection meeting, 53
    5S, *106*
agility, 154
Analyze phase, 10
attribute data, 117
    control charts, *127*
    control limit calculations, *128*
auctioning, 103–104
audience research, 113
audit worksheets, 56–57
awards and recognition, 147–148
awareness training, 51

B

balanced scorecard, 46–49
baseline assessment, 86
benchmarking, 120–121
best practices, 152–153
Black Belts, 33
brainstorming, 55

business case, 67
business process management, 159
business process reengineering, 159

C

career progression opportunities, 16, 18, 21, 51, 62, 64, 145, 155
CEO's role, 18
certification projects, 23
champion's role, 100
change, resistance to, 2
change management, 32–34, 36
check sheets, 126
civilian component, 151
coaching, 62
common cause variation, 129
common goals, 22–24
communication, 146–147
communication plans, 36, *38, 39,* 114
competing interests, 27
complexity, public sector, 28
continuous data, 117
continuous process improvement, 87, 162
control charts, 126–132
control limits, 127, *128*
Control phase, 8, 10
correlation, 138, *141*
cost reductions, 113
cost savings, 113
critical-to-quality elements (CTQ), 79–81

cross-functional teams, 85, 90
current process, 67, 87
current-state process map, *108*
customer focus, 24–25
customer satisfaction, 80
customer segmentation, 74

D

data collection, 116, 118
data stratification, 136
data transparency, 150–151
data type, 117
defect areas, 124
defect vs. defective, 127
Define phase, 7, 10
delighters, 81–82
deployment champions, 32
Design for Six Sigma (DFSS), 8, 86
diminishing returns on quality, 4
discrete data, 117
dissatisfiers, 81–82
DMADV approach, 8–9, *9*, 86
DMAIC approach, 6–8, *7, 9*

E

effective meetings and events, 113–114
ego involvement, 62
elevator pitch, 147
employee turnover, 26–27, 149–150, 151
employee types, 29–30
enterprise-wide process mapping (EWPM), 84–93
executive steering committee, 19–20, 26, 29, 32, 34–35, 149
existing data, 144
expectation setting, 31
external benchmarking, 120

F

failure to implement, 62
fear of change, 36, 61, 112
fishbone diagrams, 123–125
5 Whys technique, 123–125
5S, 102–107
flows, 90
focus groups, 76–77

funding re-allocation, 49–50
future opportunities, 153–154
future-state organization, 90
future-state process, 107, 109–112, *111*

G

goal statement, 68
goals, 44–45, 62, 67
governance structure, 151–152
Green Belts, 33
green organizations, 162–163
"ground fruit," 85
group think, 77

H

hierarchical environment, 15–16
high-level process map, *89*
high-value initiatives, 87
histograms, 132–134, 136
historical data, 118

I

implementation phobia, 62
implementation plans, 141–143
Improve phase, 10
improvement metrics, 87
in-and-out-of-scope tool, *69*
integrated communication plan, 36–37
intellectual capital, 88
internal benchmarking, 120
interviews, 13–14, 75–76
ISO standards, 159

K

kaikaku, 4
kaizen events, 30, 50–51, 100–102
Kano analysis, 81–82
keeping it simple, 51

L

leadership buy-in, 160–161
leadership performance metrics, 88
leadership support, 18–21, 61. *See also* executive steering commitee

Index   169

leadership training, 18–19
lean, 3–6
Lean-Six Sigma (LSS)
   and continuous adaptation, 160
   and "green" organizations, 162–163
   initiatives, 87
   keys to, 8
   and organizational design, 154–155
   program failure, 61–63
   public sector challenges, 13–30
   resource integration, 51
   strategy and execution, 42–44
   success components, 162
   systemic or data driven, 150
   tailored approach to, 50–51, 157
legacy processes, 23
lessons learned, 152–153

M

management support, 20
Master Black Belts, 33
Measure phase, 7–8, 10
measurement plans, 116–120
meeting goals, 113
mentoring, 62
methodology selection, 62
metric dashboard, 46–49
middle managers, 20, 160
milestones, 69
mission statements, 43
multi-voting, 59

N

negative correlation, 138, 139
negative thinking, 62
non-attribute data, 117
non-attribute environment, 55, 147, 150
non-linear correlation, *139*, 140
non-normal distribution, *134*
normal distribution, *133*

O

objective statement, 68
objectives, 44–45, 67
on-the-job training, 92

opportunities, problems and, 147
opportunity statement, 66
organizational design, 154–155
organizational turnover, 88
out-of-control indicators, 129–133
output format and data requirements, 118

P

parallel processes, 67
Pareto charts, 135–137
performance measurement, 155–156
phase tollgates, 10–11
PICK diagram, 59–61
positive correlation, 138, 140
primary defect areas, 124
problem statement, 66
process improvement, 3, 158–159
process maps, 84–93, *86*, *89*
process owners, 32
process standardization, 94–97
process steps, 92
profit focus, lack of, 21–22
program maturity, 56
project charter, 66–70
projects
   kickoff meetings, 65–66
   scope, 68–69
   selection, 22, 51–61
   stakeholders, 32–34
   teams, 63–64
   transparency, 150–151
public sector challenges, 157–158
public sector complexity, 28–29

Q

qualitative goals, 68
quantitative goals, 68
quick wins, 112

R

R chart (range), 128, *129*
rapid improvement events, 50–51
recognition, awards and, 147–148
resistance to change, 2
risk management, 40–41

S

safety phase, 105
sample size, 118, 127
satisfiers, 81–82
scatter diagrams, 138–141
scope creep, 68–69
separation of duties, 27
set-in-order phase, 104
shine phase, 104
simple tools, 51
SIPOC, 70–72, 74
Six Sigma, 4–6. *See also* Lean-Six Sigma (LSS)
SMART goals and objectives, 44–45, 67
soft skills, 148
sort phase, 103–104
sorted bar charts, 136
spaghetti diagrams, 97–99
special cause variation, 129
specification limits, 127
stakeholder analysis, 36–37, *39*
stakeholder involvement, 145–146
stakeholders, 67
standard operating procedures, 90, 94–97
standardize phase, 104
stovepipes, 16
strategic planning, 42–44
sub-processes, 89
subject matter experts, 34, 90
surveys, 77–79
sustain phase, 105
sustainment plans, 92, 143–144
SWOT analysis, 52

T

target sample size, 118
task-level detail, 90
team building, 30, 63–64, 100
team members, 32–34
theory of constraints, 159
tollgate review, 10–11

robust survey approach, 77–79
roles and responsibilities, 63–64, 150
run charts, 121–123

tool selection, 62, 63, 71, 101, 115–120, 119. *See also* individual tools
measurement, 116–120
training, 35, 51, 62, 88
transparency, project and data, 150–151
trend data, 122
TRIZ, 159
true root causes, 123
turnover, employee, 26–27, 149–150, 151

U

unrealistic goals, 62
urgency, lack of, 16–17

V

value stream, 92
value stream analysis, *110*
variation, 123
visibility, 146
visible balanced scorecard, 46–49
vision statements, 43
Vo"X" (voices)
  critical-to-quality elements, 79–81
  customer segmentation, 74
  data gathering, 75–79
  integrated approach, 83–84
  Kano analysis, 81–83
  types, 72–73

W

waste, 3–4, 25, 92, *93–94*, 97–99
what's in it for me, 16
White Belts, 33

X

X-bar (average) control chart, 128, *129*

Y-Z

Yellow Belts, 33

Belong to the Quality Community!

Established in 1946, ASQ is a global community of quality experts in all fields and industries. ASQ is dedicated to the promotion and advancement of quality tools, principles, and practices in the workplace and in the community.

The Society also serves as an advocate for quality. Its members have informed and advised the U.S. Congress, government agencies, state legislatures, and other groups and individuals worldwide on quality-related topics.

Vision

By making quality a global priority, an organizational imperative, and a personal ethic, ASQ becomes the community of choice for everyone who seeks quality technology, concepts, or tools to improve themselves and their world.

ASQ is...

- More than 90,000 individuals and 700 companies in more than 100 countries
- The world's largest organization dedicated to promoting quality
- A community of professionals striving to bring quality to their work and their lives
- The administrator of the Malcolm Baldrige National Quality Award
- A supporter of quality in all sectors including manufacturing, service, healthcare, government, and education
- YOU

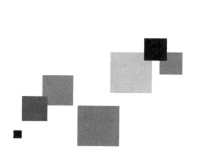

Visit www.asq.org for more information.

ASQ Membership

Research shows that people who join associations experience increased job satisfaction, earn more, and are generally happier.* ASQ membership can help you achieve this while providing the tools you need to be successful in your industry and to distinguish yourself from your competition. So why wouldn't you want to be a part of ASQ?

Networking

Have the opportunity to meet, communicate, and collaborate with your peers within the quality community through conferences and local ASQ section meetings, ASQ forums or divisions, ASQ Communities of Quality discussion boards, and more.

Professional Development

Access a wide variety of professional development tools such as books, training, and certifications at a discounted price. Also, ASQ certifications and the ASQ Career Center help enhance your quality knowledge and take your career to the next level.

Solutions

Find answers to all your quality problems, big and small, with ASQ's Knowledge Center, mentoring program, various e-newsletters, *Quality Progress* magazine, and industry-specific products.

Access to Information

Learn classic and current quality principles and theories in ASQ's Quality Information Center (QIC), *ASQ Weekly* e-newsletter, and product offerings.

Advocacy Programs

ASQ helps create a better community, government, and world through initiatives that include social responsibility, Washington advocacy, and Community Good Works.

Visit www.asq.org/membership for more information on ASQ membership.

*2008, The William E. Smith Institute for Association Research

ASQ Certification

ASQ certification is formal recognition by ASQ that an individual has demonstrated a proficiency within, and comprehension of, a specified body of knowledge at a point in time. Nearly 150,000 certifications have been issued. ASQ has members in more than 100 countries, in all industries, and in all cultures. ASQ certification is internationally accepted and recognized.

Benefits to the Individual

- New skills gained and proficiency upgraded
- Investment in your career
- Mark of technical excellence
- Assurance that you are current with emerging technologies
- Discriminator in the marketplace
- Certified professionals earn more than their uncertified counterparts
- Certification is endorsed by more than 125 companies

Benefits to the Organization

- Investment in the company's future
- Certified individuals can perfect and share new techniques in the workplace
- Certified staff are knowledgeable and able to assure product and service quality

Quality is a global concept. It spans borders, cultures, and languages. No matter what country your customers live in or what language they speak, they demand quality products and services. You and your organization also benefit from quality tools and practices. Acquire the knowledge to position yourself and your organization ahead of your competition.

Certifications Include

- Biomedical Auditor – CBA
- Calibration Technician – CCT
- HACCP Auditor – CHA
- Pharmaceutical GMP Professional – CPGP
- Quality Inspector – CQI
- Quality Auditor – CQA
- Quality Engineer – CQE
- Quality Improvement Associate – CQIA
- Quality Technician – CQT
- Quality Process Analyst – CQPA
- Reliability Engineer – CRE
- Six Sigma Black Belt – CSSBB
- Six Sigma Green Belt – CSSGB
- Software Quality Engineer – CSQE
- Manager of Quality/Organizational Excellence – CMQ/OE

Visit www.asq.org/certification to apply today!

ASQ Training

Classroom-based Training

ASQ offers training in a traditional classroom setting on a variety of topics. Our instructors are quality experts and lead courses that range from one day to four weeks, in several different cities. Classroom-based training is designed to improve quality and your organization's bottom line. Benefit from quality experts; from comprehensive, cutting-edge information; and from peers eager to share their experiences.

Web-based Training

Virtual Courses

ASQ's virtual courses provide the same expert instructors, course materials, interaction with other students, and ability to earn CEUs and RUs as our classroom-based training, without the hassle and expenses of travel. Learn in the comfort of your own home or workplace. All you need is a computer with Internet access and a telephone.

Self-paced Online Programs

These online programs allow you to work at your own pace while obtaining the quality knowledge you need. Access them whenever it is convenient for you, accommodating your schedule.

Some Training Topics Include

- Auditing
- Basic Quality
- Engineering
- Education
- Healthcare
- Government
- Food Safety
- ISO
- Leadership
- Lean
- Quality Management
- Reliability
- Six Sigma
- Social Responsibility

Visit www.asq.org/training for more information.